AF865897

Parathyroid Glands - New Aspects

Edited by Beyza Goncu and Robert Gensure

Published in London, United Kingdom

Parathyroid Glands - New Aspects
http://dx.doi.org/10.5772/intechopen.98059
Edited by Beyza Goncu and Robert Gensure

Contributors
Lívia Marcela dos Santos, Bruno Marcos Mazoca Orozco, Li-Sheng Chen, Antigoni Delantoni, Beyza Goncu, Ayça Dilruba Aslanger, Robert Gensure

© The Editor(s) and the Author(s) 2023
The rights of the editor(s) and the author(s) have been asserted in accordance with the Copyright, Designs and Patents Act 1988. All rights to the book as a whole are reserved by INTECHOPEN LIMITED. The book as a whole (compilation) cannot be reproduced, distributed or used for commercial or non-commercial purposes without INTECHOPEN LIMITED's written permission. Enquiries concerning the use of the book should be directed to INTECHOPEN LIMITED rights and permissions department (permissions@intechopen.com).
Violations are liable to prosecution under the governing Copyright Law.

(cc) BY

Individual chapters of this publication are distributed under the terms of the Creative Commons Attribution 3.0 Unported License which permits commercial use, distribution and reproduction of the individual chapters, provided the original author(s) and source publication are appropriately acknowledged. If so indicated, certain images may not be included under the Creative Commons license. In such cases users will need to obtain permission from the license holder to reproduce the material. More details and guidelines concerning content reuse and adaptation can be found at http://www.intechopen.com/copyright-policy.html.

Notice
Statements and opinions expressed in the chapters are these of the individual contributors and not necessarily those of the editors or publisher. No responsibility is accepted for the accuracy of information contained in the published chapters. The publisher assumes no responsibility for any damage or injury to persons or property arising out of the use of any materials, instructions, methods or ideas contained in the book.

First published in London, United Kingdom, 2023 by IntechOpen
IntechOpen is the global imprint of INTECHOPEN LIMITED, registered in England and Wales, registration number: 11086078, 5 Princes Gate Court, London, SW7 2QJ, United Kingdom

British Library Cataloguing-in-Publication Data
A catalogue record for this book is available from the British Library

Additional hard and PDF copies can be obtained from orders@intechopen.com

Parathyroid Glands - New Aspects
Edited by Beyza Goncu and Robert Gensure
p. cm.
Print ISBN 978-1-80355-471-6
Online ISBN 978-1-80355-472-3
eBook (PDF) ISBN 978-1-80355-473-0

We are IntechOpen, the world's leading publisher of Open Access books

Built by scientists, for scientists

6,300+
Open access books available

170,000+
International authors and editors

190M+
Downloads

156
Countries delivered to

Our authors are among the
Top 1%
most cited scientists

12.2%
Contributors from top 500 universities

BOOK CITATION INDEX
CLARIVATE ANALYTICS
INDEXED

WEB OF SCIENCE™

Selection of our books indexed in the Book Citation Index in Web of Science™ Core Collection (BKCI)

Interested in publishing with us?
Contact book.department@intechopen.com

Numbers displayed above are based on latest data collected.
For more information visit www.intechopen.com

Meet the editors

Beyza Goncu received her BS and MS in Molecular Biology and Genetics in 2007 and 2010, respectively. She obtained her Ph.D. in Biotechnology in 2018. Dr. Goncu has been involved in several scientific research projects funded by TUBITAK and the Research Funding Department of Bezmialem Vakif University (BVU), Turkey. She has published numerous journal papers, book chapters, and conference proceedings. She is a member of the European Society for Organ Transplantation (ESOT), Middle East Society for Organ Transplantation (MESOT), and the Association of Transplantation Immunology and Genetics (TIGED). Her main research area is parathyroid gland cell biology and parathyroid tissue-specific immunology. She also investigates other molecular and cellular biology fields, including therapeutic approaches. Her latest study revealed a significant outcome of tissue-specific immunogenicity through clinical features of parathyroid tissue. Dr. Goncu manages the Parathyroid Transplantation Laboratory at BVU Hospital, which aims to prepare transplantable grafts for treating hypoparathyroidism patients.

Robert Gensure, MD, Ph.D., is a physician-scientist with significant experience in basic science and clinical research as well as in the clinical practice of pediatric endocrinology. He has authored forty-six publications on topics including parathyroid hormone function, vitamin D supplementation, and inherited disorders of bone and mineral metabolism. He is currently chief of pediatric endocrinology at Tufts Medical Center, Boston, MA.

Contents

Preface

Parathyroid glands have evolutionarily conserved features and unique tissue functions for health. In the absence of parathyroid glands or in case of damage, various diseases are observed under the influence of environmental and/or genetic factors. There is no known curative treatment for parathyroid-related genetic diseases. Current treatment approaches are only palliative; they target the symptoms, not the disease itself. Observed disorders encompass a wide range of conditions in medical genetics and endocrinology and some are defined as rare diseases. Although rare in most parts of the world, parathyroid diseases are more common in certain geographic areas.

Among other composite tissues, the parathyroid glands are relatively small and thus difficult to study and research. Innovations in developing technology and biomedical advances have not adequately played a role in parathyroid gland research. Studies on the biological/functional properties of the parathyroid glands, especially those on parathyroid cell biology and molecular biology, have answered many questions to date.

This book provides a comprehensive overview of parathyroid glands.

Chapter 1 introduces the topic.

Chapter 2 describes the evolution of the parathyroid hormone (PTH) assay and defines the importance of PTH measurements. The outcomes of the assay directly affect the diagnosis and prognosis of the function of the parathyroid glands. Numerous PTH assays are available for diagnostic testing, but precision is lacking when it comes to pulsatility. As such, the chapter also discusses the limitations of the current approaches for diagnosis.

Chapter 3 examines the potential role of the trabecular bone score (TBS) evaluation in primary and secondary hyperparathyroidism. This method can be used in patients with the parathyroid-related disease to predict fracture risk.

Chapter 4 highlights the function of parathyroid glands and their relation with dentistry. Considering the critical role of the parathyroid glands in bone formation, dentists should be informed about parathyroid-related disorders such as hyperparathyroidism and hypoparathyroidism.

Chapter 5 discusses genetic diseases related to calcium and phosphorus metabolism, which affect parathyroid tissue functions in a broader context. It also presents basic and clinical information on novel issues and addresses common disturbances commonly encountered in clinical practice.

Finally, Chapter 6 evaluates the parathyroid gland with cellular and molecular biological approaches. The chapter presents studies on cell biology and the effects of many expressed molecules on the pathophysiology of parathyroid glands.

The editors would like to thank the authors of this book for their efforts and patience. This book is dedicated to all patients with parathyroid disease.

Beyza Goncu
Organ Transplantation Center,
Parathyroid Transplantation Laboratory,
Bezmialem Vakif University Hospital,
Istanbul, Turkey

Vocational School of Health Services,
Department of Medical Services and Techniques,
Bezmialem Vakif University,
Istanbul, Turkey

Robert Gensure
Tufts Medical Center,
Tufts University School of Medicine,
Boston, MA, USA

Chapter 1

Introductory Chapter: A Brief Statement about Parathyroid Glands

Beyza Goncu and Robert Gensure

1. Introduction

1.1 Discovery

Almost two centuries have passed since the discovery of parathyroid glands. Many studies have been carried out on their function and effectiveness. Many known and reputable books and studies are already available by respected scientists worldwide, including information about its embryological, developmental, anatomical, functional, and clinical importance, diagnosis, related diseases, and treatment processes. During the discovery process, the naming of the parathyroid organ was determined entirely according to its location. Considering the given name, the importance of the function provided by the parathyroid gland seems relative to be less effective than the other endocrine organs. Furthermore, the public often exposes it to preliminary assumptions as an organ related to thyroid tissue. However, they have nothing in common except that they are two organs with very different functions.

Parathyroid glands release parathormone (PTH) to perform its function and regulate the metabolism of blood calcium, phosphorus, vitamin D, and magnesium in this way [1]. Common diseases of the parathyroid glands are defined as hyperparathyroidism when the organ overproduces PTH, and hypoparathyroidism, when the organ produces less PTH or lacks PTH [2]. In the absence of the organ or the case of insufficient blood supply, individuals become deprived of the PTH hormone and its regulation.

2. Related diseases

Hyperparathyroidism is observed in cases where the function of the parathyroid glands increases above the average level. Hyperparathyroidism can occur as a secondary disease as a result of another condition. The primary disease, in this case, is chronic renal failure that causes secondary hyperparathyroidism [3]. Primary disease treatment processes are the main reason for the development of hyperparathyroidism. As a palliative treatment approach, patients are advised to use calcimimetics to suppress the production of higher PTH. Individuals have to take dialysis in cases where chronic kidney disease treatment cannot be provided. Thus, years spent in dialysis lead to overproduced PTH and increased proliferating signals to the parathyroid glands [4]. Growing parathyroid glands will release more PTH and pressure the target organs, such as bones. The individuals are partially excised by surgical methods at a certain period of their lives. Partial surgery defines as subtotal parathyroidectomy, meaning three glands plus half of the fourth gland. Rarely there are supernumerary parathyroid glands in patients [5].

IntechOpen

Nevertheless, the surgeon may control the decreased PTH level during surgery (subtotal parathyroidectomy) or, if observed by scintigraphy, then may remove the extra gland.

Further, among four parathyroid glands, hyperparathyroidism is also observed without primary reason and is called primary hyperparathyroidism. One or very rarely two of the glands may overproduce PTH. After the exact diagnosis, it is possible to treat by surgical intervention [6–8]. Another type of hyperparathyroidism is paraneoplastic hyperparathyroidism, which is very rare. Parathyroid hormone-related peptide (PTH-rp) production increases over average PTH levels due to hypercalcemia [9].

Hypoparathyroidism occurs for autoimmune or idiopathic or iatrogenic reasons. Autoimmune-related parathyroid gland diseases are observed due to genetic factors [10]. In addition, it should not be forgotten that genetic factors are considered "rare diseases" when looking at parathyroid-gland-related diseases. However, the most striking part here is that there is no known treatment option for parathyroid-gland-related genetic diseases, and developmental anomalies are observed in most diseases.

Moreover, idiopathic hypoparathyroidism refers to insufficient PTH secretion unrelated to secondary or acquired reasons. Idiopathic individuals constitute the relatively most unknown disease group for hypoparathyroidism. Iatrogenic hypoparathyroidism is observed when the surgeon unintentionally damages or removes the parathyroid gland in cases such as thyroid gland operations. Whether it is genetic, idiopathic, or iatrogenic reasons, one common problem occurs: the lack of parathyroid gland function [2, 11]. Besides, the majority of hypoparathyroidism patients are individuals who have the disease due to iatrogenic reasons. Considering the purpose and regulation mechanisms of parathyroid glands, it has a significant role in the human body. If a lack of function is observed, individuals must accept a complicated process that lasts their entire lives: permanent hypoparathyroidism [12, 13]. The regimens that should be used after that diagnosis are a part of palliative treatment that is far from improving the patient's quality of life. Much information has already been shared about the number of medicines only aimed at symptomatic treatment. Side effects and secondary diseases that may develop are becoming more severe as time passes. Individuals experiencing many side effects related to medications have been reported. In cases where symptomatic treatment is insufficient, studies about the decrease in the medication's efficiency, the time-dependent increase in the amount of medication, and the incidence of observed disorders such as anxiety and depression are rapidly taking place in the literature [14–16].

2.1 Concerns

Contemporary studies offered a specific disease-characteristic questionnaire to measure disease manifestations for hypoparathyroidism patients. In 2019, Wilde et al. used an analytical empirical approach based on retrospective analysis without involving non-disease-specific questionnaires. These testing revealed major complaints include pain and cramps, gastrointestinal symptoms, depression and anxiety, neurovegetative symptoms, and loss of vitality [17]. A recent study by Bilginer et al. performed a medication adherence questionnaire (MAQ) to hypoparathyroidism patients concerning motivation and knowledge about the palliative treatment option. Observed concerns mainly involved the side effects such as nephrotoxicity for using calcium, and kidney damage, polyuria for using active vitamin D supplementation [15]. Considering its psychological effects, hypoparathyroidism, which affects the quality of life, is evaluated from a broader perspective with current studies. In a pilot study, it was even shown that the cognitive functions of hypoparathyroidism patients were weakened

[18]. More studies are urgently needed to prevent hypoparathyroidism from the very beginning. The iatrogenic causes after thyroid surgery must be reduced.

Parathyroid imaging is essential for the location and diagnosis of hyperfunctioning glands correctly. In 2021, the European Association of Nuclear Medicine (EANM) proposed a guideline about the imaging of parathyroid glands. Several approaches and techniques were presented for nuclear medicine physicians who perform parathyroid scintigraphy, single-photon emission computed tomography/computed tomography (SPECT/CT), positron emission tomography/computed tomography (PET/CT), and positron emission tomography/magnetic resonance imaging (PET/MRI). Assessing the localization of hyperfunctioning parathyroid lesions will be more accessible by this guideline [19].

3. Treatment Options

Reckoning the diseases associated with the parathyroid tissue, the treatment options are somewhat more limited. In the development of biomedical technology, two promising treatment options come forward, particularly in the treatment of hypoparathyroidism patients: the first is hormone replacement therapy [20, 21], and the second is parathyroid transplantation [22].

3.1 Hormone replacement therapy

Hormone replacement therapy provides the chance to treat the disease in a targeted way, in this sense, Natpara®, whose Phase studies have continued success for many years [23]. However, the manufacturer has recently announced that it would not continue producing due to technical problems [24]. Transcon PTH™, a prodrug product developed for a new PTH hormone therapy, also announced that it has applied to the Food and Drug Administration (FDA) for a Phase 3 study this year [25].

3.2 Parathyroid transplantation

Parathyroid transplantation is another treatment option for hypoparathyroidism, which has a 110-year history. The first transplant belonged to Brown in 1911 [26]. In the process that started after this date in history, many researchers/physicians continued to contribute to the improvement and efficiency of parathyroid transplantation [27]. Parathyroid gland transplantation is a method still used in today's transplantation processes shared by the Cleveland Clinic [28]. Among all parathyroid transplantations, the most extensive clinical series in the literature belongs to the Warsaw group from the University of Warsaw in Poland [29]. In 2017, they reported the survival results of 316 allotransplantation data. Since the early 1990s, many research projects have been added to the literature about cell isolation, graft delivery location, and follow-up parameters from the same group [29–33]. During the last five years of the parathyroid transplantation, advanced immunological transplantation criteria, including pre-op and post-follow-up processes after parathyroid transplantation, were brought to the literature by the same group from Bezmialem Vakif University in Turkiye [22, 34–43]

Parathyroid transplantation is the most effective and targeted hypoparathyroidism treatment in the literature due to limited access to hormone replacement therapy and ongoing phase studies. Long-term studies on research and parathyroid transplantation outcomes are carried out in the literature at specific intervals. Treatment options for hypoparathyroidism have certain boundaries with more specific approaches than hyperparathyroidism. On the subject of hyperparathyroidism disorders, primary hyperparathyroidism is treated with surgical intervention and is rarely seen recurrently. Secondary hyperparathyroidism is due to another primary disease, and calcimimetics are recommended to reduce the pressure on the parathyroid glands. Considering the complaints we received from patients regarding the side effects of calcimimetics, the need for more research about the formulation of pharmaceuticals reveals that necessity.

4. Conclusion

The existing literature on treatment options for parathyroid gland diseases provides promising results. Simultaneous cellular and molecular biology studies are undoubtedly necessary and have positive effects in providing diagnostic, therapeutic, and predictive options. The unique function of the parathyroid glands and the inability to adequately treat the cause of the disease illustrate the urgent need for large cohort studies to be established. Even though the collaboration between the researchers on this subject is a fading dream, the editors hope this book will inspire such cooperation among scientific circles.

Conflict of interest

The authors declare no conflict of interest.

Author details

Beyza Goncu[1,2]* and Robert Gensure[3]

1 Organ Transplantation Center, Parathyroid Transplantation Laboratory, Bezmialem Vakif University Hospital, Istanbul, Turkey

2 Vocational School of Health Services, Department of Medical Services and Techniques, Bezmialem Vakif University, Istanbul, Turkey

3 Tufts Medical Center, Tufts University School of Medicine, Boston, MA, USA

*Address all correspondence to: bgoncu@bezmialem.edu.tr and bsgoncu@gmail.com

IntechOpen

© 2023 The Author(s). Licensee IntechOpen. This chapter is distributed under the terms of the Creative Commons Attribution License (http://creativecommons.org/licenses/by/3.0), which permits unrestricted use, distribution, and reproduction in any medium, provided the original work is properly cited.

References

[1] Lemoine S, Figueres L, Bacchetta J, Frey S, Dubourg L. Calcium homeostasis and hyperparathyroidism: Nephrologic and endocrinologic points of view. Annales d'endocrinologie. 2022;**83**(4):237-243

[2] Bilezikian JP. Hypoparathyroidism. The Journal of Clinical Endocrinology and Metabolism. 2020;**105**:6

[3] Muppidi V, Meegada SR, Rehman A. Secondary Hyperparathyroidism. Treasure Island (FL): StatPearls; 2022

[4] Ketteler M, Bover J, Mazzaferro S. ERA CKD-MBD Working Groups. Treatment of secondary hyperparathyroidism in non-dialysis CKD: An appraisal 2022s. Nephrology, Dialysis, Transplantation. 17 Aug 2022:gfac236. DOI: 10.1093/ndt/gfac236. [Epub ahead of print] PMID: 35977397

[5] Pattou FN, Pellissier LC, Noel C, Wambergue F, Huglo DG, Proye CA. Supernumerary parathyroid glands: Frequency and surgical significance in treatment of renal hyperparathyroidism. World Journal of Surgery. 2000;**24**(11):1330-1334

[6] Pal R, Jignesh SR, Sanyasi S, Lohani S, Dahiya D, Bhadada SK. Revisiting captain Charles Martell: Witnessing the true face of primary hyperparathyroidism even in the twenty-first century. Archives of Osteoporosis. 2022;**17**(1):148

[7] Bandeira F, de Moura NJ, de Oliveira LB, Bilezikian J. Medical management of primary hyperparathyroidism. Archives in Endocrinology Metabolism. 2022;**66**(5):689-693

[8] das Neves MC, Santos RO, Ohe MN. Surgery for primary hyperparathyroidism. Archives of Endocrinology and Metabolism. 2022;**66**(5):678-688

[9] Nakajima K, Tamai M, Okaniwa S, Nakamura Y, Kobayashi M, Niwa T, et al. Humoral hypercalcemia associated with gastric carcinoma secreting parathyroid hormone: A case report and review of the literature. Endocrine Journal. 2013;**60**(5):557-562

[10] Mannstadt M, Cianferotti L, Gafni RI, Giusti F, Kemp EH, Koch CA, et al. Hypoparathyroidism: Genetics and diagnosis. Journal of Bone and Mineral Research. Dec 2022;**37**(12):2615-2629. DOI: 10.1002/jbmr.4667. Epub: 2022 Nov 14. PMID: 36375809

[11] Rubin MR. Recent advances in understanding and managing hypoparathyroidism. F1000Research. 2020;**2020**:9

[12] Khan AA, Koch CA, Van Uum S, Baillargeon JP, Bollerslev J, Brandi ML, et al. Standards of care for hypoparathyroidism in adults: A Canadian and international consensus. European Journal of Endocrinology. 2019;**180**(3):P1-P22

[13] Astor MC, Lovas K, Debowska A, Eriksen EF, Evang JA, Fossum C, et al. Epidemiology and health-related quality of life in hypoparathyroidism in Norway. The Journal of Clinical Endocrinology and Metabolism. 2016;**101**(8):3045-3053

[14] Stamm B, Blaschke M, Wilken L, Wilde D, Heppner C, Leha A, et al. The influence of conventional treatment on symptoms and complaints in patients with chronic postsurgical hypoparathyroidism. JBMR Plus. 2022;**6**(2):e10586

[15] Bilginer MC, Aydin C, Polat B, Faki S, Topaloglu O, Ersoy R, et al. Assessment of calcium and vitamin D medications adherence in patients with hypoparathyroidism after thyroidectomy. Archives of Osteoporosis. 2022;**17**(1):22

[16] Frey S, Figueres L, Pattou F, Le Bras M, Caillard C, Mathonnet M, et al. Impact of permanent post-thyroidectomy hypoparathyroidism on self-evaluation of quality of life and voice: Results from the national QoL-Hypopara study. Annals of Surgery. 2021;**274**(5):851-858

[17] Wilde D, Wilken L, Stamm B, Blaschke M, Heppner C, Chavanon ML, et al. The HPQ-development and first Administration of a Questionnaire for Hypoparathyroid patients. JBMR Plus. 2020;**4**(1):e10245

[18] Rubin MR, Tabacco G, Omeragic B, Majeed R, Hale C, Brickman AM. A pilot study of cognition among hypoparathyroid adults. Journal of Endocrinal Society. 2022;**6**(3):bvac002

[19] Petranovic Ovcaricek P, Giovanella L, Carrio Gasset I, Hindie E, Huellner MW, Luster M, et al. The EANM practice guidelines for parathyroid imaging. European Journal of Nuclear Medicine and Molecular Imaging. 2021;**48**(9):2801-2822

[20] Clarke BL, Vokes TJ, Bilezikian JP, Shoback DM, Lagast H, Mannstadt M. Effects of parathyroid hormone rhPTH(1-84) on phosphate homeostasis and vitamin D metabolism in hypoparathyroidism: REPLACE phase 3 study. Endocrine. 2017;**55**(1):273-282

[21] Vokes TJ, Mannstadt M, Levine MA, Clarke BL, Lakatos P, Chen K, et al. Recombinant human parathyroid hormone effect on health-related quality of life in adults with chronic hypoparathyroidism. The Journal of Clinical Endocrinology & Metabolism. 2017;**103**(2):722-731

[22] Goncu B, Salepcioglu Kaya H, Yucesan E, Ersoy YE, Akcakaya A. Graft survival effect of HLA-A allele matching parathyroid allotransplantation. Journal of Investigative Medicine. 2021;**69**(3):785-788

[23] Yao L, Li J, Li M, Lin C, Hui X, Tamilselvan D, et al. Parathyroid hormone therapy for managing chronic hypoparathyroidism: A systematic review and Meta-analysis. Journal of Bone and Mineral Research. Dec 2022;**37**(12):2654-2662. DOI: 10.1002/jbmr.4676. Epub: 2022 Nov 16. PMID: 36385517

[24] Takeda to Discontinue Manufacturing of NATPAR®/NATPARA® for Patients with Hypoparathyroidism at the End of 2024 [Internet]. Takeda Pharmaceutical Company Limited. 2022 [cited December 5, 2022]. Available from: https://www.takeda.com/en-us/newsroom/statements/2022/takeda-to-discontinue-manufacturing-of-natpar-natpara/

[25] TRANSCON PTH Designed to replace parathyroid hormone to physiologic levels in patients with hypoparathyroidism [Internet]. Ascendis Pharma. [cited December 5, 2022]. Available from: https://ascendispharma.com/pipeline/endocrinology/transcon-pth/

[26] Brown WH. Parathyroid implantation in the treatment of Tetania Parathyreopriva. Annals of Surgery. 1911;**53**(3):305-317

[27] Zhang JLH, Appelman-Dijkstra NM, SchepersA.ParathyroidAllotransplantation: A systematic review. Medical Science (Basel). 2022;**10**(1):19-32

[28] Agarwal A, Waghray A, Gupta S, Sharma R, Milas M. Cryopreservation of parathyroid tissue: An illustrated technique using the Cleveland clinic protocol. Journal of the American College of Surgeons. 2013;**216**(1):e1-e9

[29] Barczynski M, Golkowski F, Nawrot I. Parathyroid transplantation in thyroid surgery. Gland Surgery. 2017;**6**(5):530-536

[30] Nawrot I, Wozniewicz B, Szmidt J, Sladowski D, Zajac K, Chudzinski W. Xenotransplantation of human cultured parathyroid progenitor cells into mouse peritoneum does not induce rejection reaction. Central European Journal of Immunology. 2014;**39**(3):279-284

[31] Nawrot I, Woźniewicz B, Tołłoczko T, Sawicki A, Górski A, Chudziński W, et al. Allotransplantation of cultured parathyroid progenitor cells without immunosuppression: Clinical results. Transplantation. 2007;**83**(6):734-740

[32] Tolloczko T, Wozniewicz B, Gorski A, Sawicki A, Nawrot I, Migaj M, et al. Cultured parathyroid cells allotransplantation without immunosuppression for treatment of intractable hypoparathyroidism. Annals of Transplantation. 1996;**1**(1):51-53

[33] Wozniewicz B, Migaj M, Giera B, Prokurat A, Tolloczko T, Sawicki A, et al. Cell culture preparation of human parathyroid cells for allotransplantation without immunosuppression. Transplantation Proceedings. 1996;**28**(6):3542-3544

[34] Goncu B, Yucesan E, Basoglu H, Gul B, Aysan E, Ersoy YE. Xenotransplantation of microencapsulated parathyroid cells as a potential treatment for autoimmune-related hypoparathyroidism. Experimental and Clinical Transplantation. 9 Aug 2021. DOI: 10.6002/ect.2020.0403. Epub ahead of print. PMID: 34387152

[35] Goncu B, Yucesan E, Ersoy YE, Aysan ME, Ozten KN. HLA-DR, -DP, -DQ expression status of parathyroid tissue as a potential parathyroid donor selection criteria and review of literature. Human Immunology. 2021;**83**(2):113-118

[36] Aysan E, Yucesan E, Goncu B, Idiz UO. Fresh tissue parathyroid Allotransplantation from a cadaveric donor without immunosuppression: A 3-year follow-up. The American Surgeon. 2020;**86**(4):e180-e1e2

[37] Yucesan EGB, Idiz O, Ucak R, Ozdemir B, Kanimdan E, Ersoy YE, et al. The effect of intravenously paratyroid cell xenotransplantation in sheep: As an animal model. Kafkas Universitesi Veteriner Fakultesi Dergisi. 2020;**26**(6):765-770

[38] Aysan E, Yucesan E, Idiz UO, Goncu B. Discharging a patient treated with parathyroid allotransplantation after having been hospitalized for 3.5 years with permanent hypoparathyroidism: A case report. Transplantation Proceedings. 2019;**51**(9):3186-3188

[39] Yucesan E, Goncu B, Ozdemir B, Idiz O, Ersoy YE, Aysan E. Importance of HLA typing, PRA and DSA tests for successful parathyroid allotransplantation. Immunobiology. 2019;**224**(4):485-489

[40] Yucesan E, Basoglu H, Goncu B, Akbas F, Ersoy YE, Aysan E. Microencapsulated parathyroid allotransplantation in the omental tissue. Artificial Organs. 2019;**43**(10):1022-1027

[41] Goncu B, Yucesan E, Aysan E, Kandas NO. HLA class I expression changes in different types of cultured parathyroid cells. Experimental and Clinical Transplantation. 2019

[42] Goncu B, Yucesan E, Ozdemir B, Basoglu H, Kandas NO, Akbas F, et al. A new transport solution for parathyroid allotransplantation: Effects on cell viability and calcium-sensing receptors. Biopreservation and Biobanking. 2018;**16**(4):278-284

[43] Yucesan E, Goncu B, Basoglu H, Ozten Kandas N, Ersoy YE, Akbas F, et al. Fresh tissue parathyroid allotransplantation with short-term immunosuppression: 1-year follow-up. Clinical Transplantation. 2017;**31**(11):10-14

Chapter 2

PTH Measurement in Clinical Laboratories

Li-Sheng Chen

Abstract

In this chapter, we will start with a review of the methodological evolution of the clinical parathyroid hormone (PTH) assays, follow with a detailed discussion of clinical utility, analytical and clinical performances of the current second and third generation assays, their drawbacks and the efforts taken collaboratively by academia and industry to harmonize the PTH assays. Next, we will focus on the profiling of various forms of circulating PTH in healthy and diseases by LC-MS/MS-based analysis, which greatly contribute to the advancement of our understanding in the structure/function and pathophysiology of PTH over the past three decades. Finally, we will comment on the remaining challenges of the present PTH assays for patient management and point to the future research and development needs to meet the unmet medical needs in managing patients with hyperparathyroidism and chronic kidney diseases–mineral and bone disorder (CKD-MBD).

Keywords: parathyroid hormone, intraoperative PTH assay, secondary hyperparathyroidism, chronic kidney diseases

1. Introduction

Calcium and phosphate are critical to skeletal mineralization; while ionized calcium is essential for neuromuscular function and serves as a signaling molecule to communicate and drive intracellular processes. Although, only about 1% of total body calcium and 15% of total body phosphorus is in circulation, the ionized fractions of circulating calcium and phosphate are tightly regulated by the interplay of several hormones to keep their status of homeostasis in response to environmental cues and the physiological needs [1].

PTH and 1,25-dihydroxy vitamin D are the major regulators of calcium metabolism, while PTH and EGF 23 and its cofactor, klotho, work concertedly to control renal excretion of phosphorus and maintain phosphate balance. The complex system is at play to keep these hormones in check in the healthy, while this intricate control mechanism is disrupted in diseases due to the hormone excess/deficiency or loss of the metabolite feedback control such as in patients with parathyroid gland dysfunction or chronic kidney diseases. Therefore, timely and accurately assessing, monitoring, and profiling of these hormones and the important metabolites is essential for the clinicians to understand the degree of calcium and phosphate imbalance when they evaluate the related disorders such as hypoparathyroidism, various forms of hyperparathyroidism, and chronic kidney disease-induced mineral and bone disorder (CKD-MBD) [2, 3].

PTH measurement has been used in the diagnosis and treatment of disorders of calcium/phosphate metabolism because of its predominant role in maintaining the circulating ionized calcium within a very tight concentration range, and in regulating the urinary excretion of phosphorus. PTH measurement is a valuable tool for diagnosing primary and secondary hyperparathyroidism (SHPT). It is also used as a surrogate biomarker to guide the management strategies for CKD patients presenting with systemic mineral and bone disorders (CKD-MBD) and monitoring its progression.

The presence of various circulating forms of PTH and its metabolites, the inter-assay variability and the presence of many variables from sample collection to the test reporting pose significant challenges for accurate PTH quantitation in clinical laboratories and the interpretation of PTH results by clinicians.

2. Evolution of PTH assays

Circulating PTH is a heterogeneous population consisting of full-length PTH (84 amino acids, with a molecular of ~9500 Da) and various sizes of proteolytic C-terminal, N-terminal, and mid-molecule metabolites [4]. In healthy individuals, predominant C-terminal PTH fragments typically started at amino acid position 34, 37, 38, or 45 [5]; a subtype of C-PTH, known as non-1-84 or PTH (7-84), usually starts at amino acid position 4, 7, 8, 10, or 15 with the major fragment presumably starting at position 7 [6, 7]. Full-length PTH (1-84) and N-terminal PTH fragments have very short half-lives (2–4 min), while the C-terminal PTH fragments have a half-life of several hours and even longer in CKD patients with decreased renal clearance [8]. The half-life of PTH (7-84) fragments is longer, ranging from 8.1 to 24.0 min [9]. It was shown that PTH (7-84) fragments are released from the parathyroid gland directly in healthy individuals, but proportionally increase relative to total circulating PTH due to bioaccumulation in patients with CKD [10].

The first C-terminal PTH radioimmunoassay was described by Berson et al. [11]; this and the subsequent first-generation PTH assays employed a single polyclonal antibody against epitopes that were located within the C-terminal part of PTH and thus detected both PTH (1-84) and all C-PTH fragments. It was found that PTH (1-84) and C-terminal PTH fragments accounted for 20% and 80% of the circulating PTH respectively in healthy adults when measured by the first-generation PTH assay [12]. Meanwhile, in CKD patients, the proportion of measured C-terminal PTH fragments increased to 95% of circulating PTH [13]. The first generation PTH RIA assay is time-consuming and lacks specificity, especially in CKD patients and thus was totally replaced by more specific second-generation sandwich assays.

Nichols Diagnostics developed a two-site immunoradiometric assay (IRMA) for measuring PTH in 1987; this assay uses a capture antibody directed against the 39-84 C-terminal epitope region and a signaling antibody directed towards the 13–4 N-terminal epitope region to form antibody-PTH-antibody complex and thus greatly improved sensitivity and specificity of the PTH quantitation [14].

Subsequently, non-radioactive labeling assay formats (ELISA, chemiluminescent, and electrochemiluminescent methods) were brought forth and operated in automated immunoanalyzers in clinical laboratories of all sizes, which became and still remain to be the most widely-used PTH assays to date.

These second-generation assays were collectively known as "intact" PTH assays, for it was thought that they measure only the full-length PTH (1-84). However, it was uncovered later that the "intact" PTH assays still cross-react with PTH (7-84) fragments (ranging from 50% to 100%), and thus overestimated PTH concentration in CKD patients [15].

To improve the diagnostic accuracy, the first third-generation "biointact," (also known as "bioactive" or "whole") PTH (1-84) assay, was advanced by Scantibodies Laboratories and became available as an exoteric testing service since 1999. It is an immunochemiluminometric assay with a signaling antibody directed against the epitope within the first 4 amino acids at the very N-terminus of full-length PTH [16]. More recently, non-radioactive automated and FDA-cleared third-generation assays were marketed by several manufacturers including DiaSorin, Tosoh, Fujirebio, Roche, and bioM´erieux [17]. Since third-generation assays have higher specificity to PTH (1-84) and won't cross-react with C-PTH, the measuring values are approximately 50–70% of those measured by the second generation PTH assay in patients with CKD and approximately 15% lower than those in persons without CKD. Because second generations assays have been used for decades and are still widely in use, this inter-generation assay difference complicates the test interpretation and the adoption of the new assays. It was expected that the use of the third-generation assays can resolve the issue of cross-reactivity with non-functional PTH fragments (i.e., PTH 7-84). However, subsequent studies revealed the complexity of PTH physiology that was undetected by the older PTH assays. In addition to the full-length PTH (1-84), the third-generation "biointact" PTH also reacted with a new form of N-terminal PTH (N-PTH) that is not recognized by most second-generation PTH assays. Further investigation showed that posttranslational phosphorylation at serine position 17 of the PTH (1-84) molecule prevent (or reduces the binding affinity of) the signaling antibody in most second-generation assays from binding to its epitope in the phosphorylated N-PTH. N-PTH accounts for 4–8% versus 15% of circulating PTH measured by the third generation PTH assays in healthy versus in patients with CKD [16, 18].

3. Unresolved problems with PTH assays

3.1 Inter-assay variability

Currently, automated second-generation and third-generation PTH assays are employed in clinical laboratories for PTH measurement. Current second-generation assays provide convenient and relatively reliable methods (intra-assay imprecision <10 %) for PTH measurement [19]. However, different PTH assays from various assay manufacturers measure different types and amounts of the circulating PTH forms depending on the specificity of the antibodies used to construct the assay, which led to great inter-assay variability and inconsistent results among the PTH measurements when the now-obsolete Allegro PTH intact assay served as the reference [20–22]. In a more recent study, a performance comparison among six currently-existing second-generation assays was made. Imprecision was evaluated using three concentrations of commercial quality control materials, while inter-assay variability was assessed by paired comparisons using 203 serum and 193 EDTA plasma samples from healthy individuals. The results showed that the imprecision (i.e., total coefficients of variation) were between 1.1% and 10.9% and there was a good correlation for all methods overall but the considerable bias was observed between methods, the

Bland-Altman plots revealed that the between assay differences were between +1.6% to −36.3%, influenced by both assays and sample types used [23].

The results of third-generation PTH assays are approximately 50–70% of those measured by the intact PTH assay in patients with CKD and approximately 15% lower than those in persons without CKD [16, 18]. The automated third-generation assays are calibrated against the WHO 95/646 Standard and therefore displayed significantly improved inter-method agreements [24, 25]. However, the incompatibility in measurement to that of the widely-used second generation assays affect the interpretations of the method validation and may contribute to its slow adaptation to clinical laboratories in general.

3.2 Aggravating heterogeneity of circulating PTH in the disease states

The pathological changes in calcium and phosphate status and the progressive loss of feedback control in the calcium and phosphate regulatory system in hemodialyzed patients further exacerbate the problem of assay variability. A systematic performance evaluation of 15 commercial immunoassays with 47 serum pools from dialysis patients, reported by Souberbielle et al. [21] in 2006, showed great inter-assay variability among the tested PTH assays, moreover, the discrepancies of measured values in some assays compared to the then "gold-standard" Allegro PTH intact assay are unacceptable, and may cause patient harm when the discrepant results were used to make therapeutic decisions. This raised the alarm in the dialysis community to question the reliability of the PTH testing and resulted in many more investigations on these critical issues. A recent position paper issued by the IFCC Committee for Bone Metabolism tabulated 23 major assay comparison studies using samples from CKD or hemodialysis patients (published between 2005 and 2018) [17]. The results reaffirmed that existing between-method differences in PTH measurements did not improve much and likely have treatment implications.

As described earlier, third-generation assays cross-react with a phosphorylated form N-PTH was overproduced in some patients with parathyroid carcinoma and severe primary hyperparathyroidism (PHPT) [26, 27]. In these cases, PTH determination with the third generation assay will have a value greater than the one with the second-generation assay. The inverted third/second PTH ratio is therefore proposed as a screening or monitoring tool for parathyroid cancer [28]. However, the inter-assay variability makes it challenging to define a generally acceptable cutoff for validating the proposed clinical utility unless the problem of analytical variability is effectively addressed.

3.3 Issues of concern related to testing procedures

3.3.1 Pre-analytical phase: sample stability, sample type, and sampling time

The unstable nature of PTH makes it essential to optimize pre-analytical parameters, including specimen type, sampling time, and storage conditions, which have all been thoroughly investigated. After a systematic review conducted under the auspice of IFCC PTH Working Group, the following evidence-based recommendations are made by IFCC: [29, 30].

- For samples collected with EDTA tubes, the plasma must be separated from the cells within 24 hours of venipuncture. Samples should be kept at 4°C and analyzed within 72 hours of venipuncture.

- For serum samples, the serum must be separated from the cells as soon as possible, and PTH is analyzed within 3-4 hours of venipuncture.

- Central venous PTH concentrations were higher compared to peripheral venous PTH concentration, therefore, in patients undergoing hemodialysis or parathyroidectomy, if the blood samples were collected via central line or central vein; the collected tube, as well as the test report, should explicitly state the collection site and whether they are peripheral or central venous concentrations.

- PTH follows a circadian rhythm, exhibiting a nocturnal peak, a mid-morning nadir, and a smaller afternoon peak. Therefore, it is suggested that samples for PTH measurement should be collected between 10:00 and 16:00, preferably in the morning with an overnight fast. Other known biological variations include the fact that PTH level increases with age and BMI, and is generally higher in African Americans than in Caucasians.

PTH has longer stability in EDTA tube at room temperature than in serum tube, thus delayed centrifugation to allow blood clotting is not needed; however, it is important in clinical practice that PTH measurement be accompanied by a concomitant calcium value. Since calcium (and bone-alkaline phosphatase) cannot be measured in EDTA plasma, PTH and calcium are to be measured in the same serum tube and therefore may be a preferred option for practical reasons. In two recent reports, PTH values obtained from the rapid serum tubes were found to be decreased compared to those from the serum separator tubes (SSTs) [31, 32].

So far, there is no reported data on the comparison of relative PTH stability using second-generation versus third-generation assays. Such study is valuable in providing further insight into the *ex vivo* stability/vulnerability of each type of PTH molecule.

3.3.2 Analytical phase: heterophilic antibodies interference

All immunometric assays are inherently prone to interference from heterophilic antibodies (i.e., human idiotypic antibodies that interact with assay antibodies raised from animals). Such interference can lead to diagnostic errors and may cause harm to patients as consequence. Assay manufacturers have introduced effective blockers to the assay reagent as a preventive measure against heterophilic antibody interference; however, increasing use of modified monoclonal mouse antibodies as therapeutics in recent years makes heterophilic antibodies interference a special concern in the patients who receive such treatment. Clinicians and laboratorians should keep open communication when the testing results did not match the clinical picture of the patients. Laboratorians should offer adequate confirmatory measures and be able to interpret the investigative results timely and correctly to avoid the spurious results being used to make important clinical decisions for patient management [33].

3.3.3 Post-analytical phase: reference ranges

Since a great inter-assay variability is still present among current commercial PTH assays, there also exists a significant difference for the reference ranges provided by the manufacturers. This makes validation of the PTH assay in use an indispensable but tedious and challenging job. The first and foremost is the selection of the reference population; The eGFR, serum calcium, and 25[OH]D values of the candidate samples should be determined to only include those within reference ranges, especially the 25[OH]D level should be >30 ng/mL [34, 35]. The reference ranges should be established for each sample locally used. If types of collection tubes differ among the collection sites, it is advisable to perform a comparison study to determine if the reference range should be revised to accommodate the difference. In high latitude areas, the effect of seasonal variation in 25[OH]D levels may need to be taken into consideration in designing the validation study for establishing the local reference range for PTH assay.

3.4 Inability to discern oxidized from non-oxidized PTH by current clinical PTH assays

Loss of biological activity of oxidized PTH *in vitro* was observed as early as 1934, it was later showed that oxidation of two methionine residues at positions 8 and/or 18 within the receptor-binding domain results in the altered three-dimensional conformation of PTH, which in turn results in greatly reduced affinity to PTH receptor [36]. Other *in vitro* studies also provide evidence that oxidized PTH does not stimulate cAMP-mediated signal after binding to its cellular receptor and thus lost the ability to activate muscle contraction, nor can it stimulate mouse calvarial bone cells to increase alkaline phosphatase activity, which indicates that excessive oxidative stress may disrupt calcium and phosphate homeostasis through oxidation of PTH [37].

There is plenty of experimental evidence to support the increase of oxidative stress in patients with CKD due both to the depletion of anti-oxidants and increase of reactive oxidative species (ROS) production; increasing oxidative stress is also shown to be associated with complications such as hypertension, atherosclerosis, and anemia and therefore may contribute to the accelerated disease progression and mortality in CKD [38]. However, none of the current second-and third-generation assays can discern oxidized from non-oxidized PTH until recently due to the lack of appropriate analytical tools for investigation.

Hocher et al. developed a two-step method to measure the non-oxidized PTH; the oxidized PTH molecules were first removed by an immunoaffinity column with monoclonal antibodies specifically against the oxidized human PTH (1–34) fragment, the remaining non-oxidized PTH was then measured by Roche second-generation PTH assay. The method was applied to analyze 17 hemodialyzed samples and revealed a substantial but variable portion (70–90%) of the total PTH was in oxidized form for all samples tested [39]. These tools enable the researchers to assess iPTH, non-oxidized PTH and oxidized PTH simultaneously using the same parameters in clinical research settings to address the clinical association of oxidized PTH with the progression of CKD [40, 41]. However, the results from these studies are not conclusive.

The inherent problems of the analytical procedure are a concern. First, the *ex vivo* oxidation after sample collection cannot be ruled out; second, the recovery of non-oxidized PTH after the immunoaffinity column removal of oxidized PTH is unknown in clinical samples, and most troubling of all, iPTH assays has not been standardized yet, it is known that some of the iPTH assays use a signaling antibody that is raised against the epitope close to the second oxidation site (methionine at position 18) in PTH; the avidity of such signaling antibody may be changed by the oxidation of PTH. Therefore, replacing second-generation intact PTH with a more specific third-generation PTH (1-84) assay, introducing the spiked internal control to calculate recovery after column treatment, and devising standard operating procedures to minimize and evaluate the extent of *ex vivo* oxidation will improve the reliability of this assay. For now, the non-oxidized PTH is not ready for clinical use unless all the issues described above are appropriately addressed.

3.5 Clinical implications of the problematic PTH measurements

Lack of a common PTH reference range not merely cause inconvenience, but it also affects results interpretation, and possibly clinical management, especially for monitoring long-term changes of PTH level if patients are not able to use the same health care facilities. More importantly, intraindividual biological variability of PTH is known to increase in hemodialysis patients. The negative impact of PTH assay variability on the management of patients with CKD and hyperparathyroidism is especially troublesome.

3.5.1 Clinical practice guidelines for PTH measurement in CKD-MBD

In 2003, National Kidney Foundation—Kidney Disease Outcomes Quality Initiative (KDOQI) published a guideline that recommended maintaining a target range of 150–300 pg/mL for intact PTH concentrations in stage 5 CKD patients to reduce the mortality related to CKD-MBD. However, the recommendation was based on the comparison of PTH measurements using Allegro iPTH (now obsolete) with the gold standard–bone biopsy, before the problem of inter-assay variability being revealed. A later study showed that iPTH (measured by Immulite DCP assay) levels less than 150 pg/ml for identifying low turnover and greater than 300 pg/ml for high turnover presented a positive predictive value of 83% and 62%, respectively, moreover, in patients achieving the target iPTH levels, 88% had low turnover diseases [42]. The great inter-assay variability is a likely contributing factor to the poor results and indicated the misclassifications may cause harmful clinical outcomes [21].

In one study conducted by the United Kingdom National External Quality Assessment Service (UK-NEQAS), a 4.2-fold difference between highest and lowest measured PTH concentrations were observed using five commonly-used second-generation assays when testing EDTA plasma from 21 hemodialysis patients. In a subsequent study, 98 patient samples were tested by the same iPTH assays to derive assay-specific target values based on Passing and Bablok regression against Roche Elecsys E170 assay which gave the closest results to target values recommended by clinical guidelines. By applying the corrected assay-specific target values, the misclassifications of bone turnover reduced from 53% to 12% [43].

The Kidney Disease Improving Global Outcomes (KDIGO) 2009 Guidelines for the Diagnosis, Evaluation, Prevention and Treatment of CKD-MBD expanded the scope and refined the recommendations to assist clinicians in treating patients with CKD Stages 3–5 who are on dialysis. Aware of the problem of inter-assay variability, this guideline avoids the use of absolute PTH values but suggests hemodialyzed patients maintain PTH levels between two and nine times the upper normal limit (ULN) of the assay used and emphasize on trending the changing pattern rather than the value per se. PTH values above the target suggest high bone turnover bone disease with a specificity of 86%, while PTH levels below the target value suggest low bone turnover with a sensitivity of 66%. In its recent update (July 2017), targets for CKD-MBD biomarkers, including PTH, remain unchanged [44]. As mentioned earlier, 25[OH]D status has a substantial influence on physiological PTH level, but PTH assay reference ranges offered by the vendors might be established with samples of poor 25[OH]D status. Recent studies reported that the upper reference range for PTH established with a reference population with normal GFR and calcium levels, and a 25[OH]D level >20 ng/mL is significantly lower than the reference range provided by the manufacturers [35, 45, 46]. Cavalier et al. found that applying KDIGO guideline PTH target ranges using reference ranges established in a vitamin D replete healthy control population would reduce the percentage of misclassification of bone turnover in dialyzed patients to 16% (versus 36% using vendor-established reference range) [34].

3.5.2 PTH assay incompatibility and the diagnosis of PHPT

Comparison studies compared Nichols iPTH versus Bio-intact PTH assays and Scantibodies Laboratory's Total versus Whole PTH assays showed high diagnostic sensitivity for both second-and third-generation PTH assays (89–97%), which provided evidence that both types of PTH assays are valuable tools in diagnosing PHPT and provide comparable results [47].

It is essential, as described earlier, to use a vitamin D-replete population to establish the reference range for PTH; the diagnostic accuracy (sensitivity and specificity) for PHPT was shown to be improved in a vitamin D-replete population [46, 48]. However, there is debate over using sufficiency level (30 ng/mL) or insufficiency level (20 ng/mL) as the threshold. So far, there are still no established reference intervals for second-and third-generation PTH assays using large vitamin D-replete population cohorts; subjects with hypercalcemia and a PTH persistently within the upper reference range should be considered "asymptomatic" PHPT and closely monitored [49].

3.5.3 Assay-dependent rate change in intra-operative PTH testing

Because PTH (1-84) has a very short half-life (~5 min), measuring PTH concentrations during parathyroidectomy can inform surgeons whether the pathological parathyroid tissue has been removed completely. A 50% decline of PTH compared with the preoperative level is commonly used to define treatment success. Since second-generation assays cross-react with C-terminal fragments with a longer half-life, it is reasonable to think third-generation assays will perform better for intraoperative PTH monitoring. Studies so far did not demonstrate that third-generation assays have a better performance for ioPTH in patients with PHPT. But a more rapid rate of PTH drop was observed in the third-generation assays. In surgery performed in patients with SHPT, it takes time for PTH concentrations to drop below the 50% cutoff after removal of the last hyperplastic gland using second-generation assays. More studies are needed to determine if third-generation assays offer a superior clinical utility for ioPTH monitoring, especially in patients with SHPT [50].

In addition, PTH measurement is used 20 min after thyroidectomy to determine if intensive calcium monitoring is needed (when the PTH level is >15 pg/mL using iPTH assays) and 4 hours after to predict postoperative hypocalcemia [51, 52]. There is no published study comparing the performance of different-generation PTH assays for this particular purpose, although one can argue that third-generation PTH measurement may better reflect the biological activity of parathyroid glands.

4. Toward standardization of PTH assays

There is a desperate need to reduce the significant inter-assay variability of PTH measurement that confounds the test interpretation and may affect clinical decision-making and cause patient harm. Assay standardization is therefore urgently required to improve the long-standing troubling situation in clinical PTH testing. Hormonal immunoassay standardization is inherently challenging because of the low circulating concentration, protein instability, and the differences in antibody-antigen complex formations; the presence of various PTH fragments and their different distributions in response to calcium, 25(OH) D status, and other effectors present additional hurdles to overcome.

International Federation of Clinical Chemistry and Laboratory Medicine (IFCC), equipped with the experience for standardizing thyroid hormone and other protein assays, is undertaking the challenge to improve the PTH measurement in clinical communities. The IFCC Committee on Bone Metabolism has laid out the roadmap in a position statement where they set three major priorities:

- Calibrate all current commercial PTH assays against a recognized International Standard, proposed to be the recombinant human PTH 1-84 standard (NIBSC 95/646) prepared by World Health Organization.

- Facilitate the development of reference measurement procedure (RMP) for PTH (1-84) to enable the true metrological assignment of reference value for PTH primary, secondary and working standards in a network of reference laboratories.

- Design studies to establish common reference intervals for PTH assays [30].

4.1 Assessing assay commutability and establishing reference PTH sample panels

The first task for the IFCC Working Group is to assess the commutability of PTH in a defined matrix, which is to demonstrate experimentally that the standard material and fresh patient specimens exhibit the same analytical response (regression line slope close to 1.0) when measured by two different methods.

A collaborative effort is currently undertaken to develop a protocol for the formal assessment of commutability. Once the commutability of the standard material is determined and deemed acceptable, it will be possible to use RMP to determine the recovery of PTH (1-84) in the appropriate matrices and then certify the values for secondary reference materials and external controls. Once such standards are available, assay manufacturers should use them to calibrate their assays.

In the meantime, the work is also underway to acquire an appropriate panel of plasma and/or serum samples for establishing PTH reference intervals. All the pre-analytical and physiological factors that can contribute to intra- and inter-individual variations as well as to increase inter-assay variability, as discussed in the previous section of this chapter, are to be carefully considered and minimized or ruled out.

One obvious challenge for this endeavor is the lack of consensus on how to define vitamin D sufficiency, insufficiency, and optimal vitamin D levels. The most often used definition of vitamin D sufficiency is the 25[OH]D concentration above which PTH cannot be suppressed further, however, this threshold varies with disease states and is subject to analytical variability of 25-(OH) D assays [39]. Recently, a cross-sectional analysis of 14,289 CKD patients (stages 1–5) and a randomized control trial involving 429 patients with stage 3-4 CKD, showed levels of iPTH was not suppressed until serum 25(OH) D reached 40–50 ng/ml range, therefore, the target 25-(OH) D concentration may need to be raised in CKD patients [53, 54].

4.2 Developing candidate reference PTH measurement procedures for assay standardization

The liquid chromatography-tandem mass spectrometric (LC-MS/MS) method utilizes chromatographical separation and distinct mass/charge ratio of the product ion pairs to provide rigorous physicochemical characterization of target molecules in the biological mixture and therefore is ideal for developing reference measurement procedures for PTH. The technical advances in mass spectrometric analysis in the last decade enable LC-MS/MS to quantify PTH with accuracy and precision comparable to the results obtained with immunoassays in complex matrices, while it is more robust and flexible for identifying and measuring new or modified PTH fragments (i.e., oxidized PTH). Several published LC-MS/MS methods for PTH measurement already exist and can be readily refined and modified to become candidates for RMPs [5, 55].

IFCC working group has conducted a feasibility study as the first step to assess the suitability of a selected LC-MS/MS method as RMP for PTH quantification. In this study, 48 freeze-dried proficiency testing specimens with assigned values sent from UK NEQAS were reconstituted and analyzed by a published LC-MS/MS method used at the Mayo Clinic [56]. Results obtained from LC-MS/MS analysis were in excellent agreement with the target all laboratory trimmed mean used in the UK NEQAS for PTH and thus the feasibility of using the LC-MS/MS method as a candidate reference measurement procedure.

However, the analytical sensitivity of current MS methods still could not match that provided by immunoassays for PTH quantitation; moreover, current methods require proteolytic digestion of PTH before MS analysis, which is time-consuming and can introduce significant procedural variability. Some of the MS methods also include an immunoabsorbent step to select and enrich PTH. The specificity of the antibody used will influence what types of PTH fragments later be analyzed by LC-MS/MS and thus introduce biases to the final results. Therefore, there remain many hurdles to overcome in developing an MS-based RMP for PTH measurement [30].

Moreover, the employment of state-of-art liquid chromatography-high resolution mass spectrometry (LC-HRMS) could potentially profile various PTH fragments in different stages of CKD with a sensitivity comparable to that by immunoassays and thus offer a powerful tool to correlate PTH qualitative and quantitative changes with the progression of CKD [57].

5. Conclusions

Accurate quantitation of circulating PTH is challenging due to the presence of various molecular forms of PTH and its complex physiological interactions with other hormones and its effectors—calcium, and phosphate. PTH assays have been continuously evolving and improving since their debut six decades ago; the effort will continue to refine and adapt to resolve the issues at hand and meet the evolving clinical needs.

DOI: http://dx.doi.org/10.5772/intechopen.103894

To improve the reliability of the PTH testing for diagnosis and monitoring along the pathway of patient management, IFCC has spearheaded an ambitious plan to standardize commercial PTH assays. It will require the collaborative efforts of academics, scientific and clinical communities, assay manufacturers, and the support of other stakeholders to achieve the goals. We can be optimistically hopeful that the communicable PTH reference material, the panels of qualified samples for establishing reference ranges, and the LC-MS/MS-based RMP method(s) will be available in the foreseeable future. Together they will enable calibration of all PTH assays with a single reliable international standard and allow accuracy-based external quality assessment. The better analytical tool can also empower us to gain more insight into the dynamic changes of PTH so we can optimize the testing to be used in the management of CKD to benefit patient care.

Author details

Li-Sheng Chen
Independent Researcher, Silver Spring, Maryland, USA

*Address all correspondence to: lishengchen66@gmail.com

IntechOpen

© 2022 The Author(s). Licensee IntechOpen. This chapter is distributed under the terms of the Creative Commons Attribution License (http://creativecommons.org/licenses/by/3.0), which permits unrestricted use, distribution, and reproduction in any medium, provided the original work is properly cited.

References

[1] Fraser WD. Bone and mineral metabolism. In: Tietz Textbook of Clinical Chemistry and Molecular Diagnostics. 6th ed. Vol. 64. St. Louis, Missouri: Elsevier; 2018. pp. 1422-1491

[2] Bergwitz C, Jüppner H. Regulation of phosphate homeostasis by PTH, vitamin D, and FGF23. Annual Review of Medicine. 2010;**61**:91-104

[3] Peacock M. Phosphate metabolism in health and disease. Calcified Tissue International. 2021;**108**(1):3-15

[4] Habener JF, Segre GV, Powell D, Murray TM, Potts JT. Immunoreactive parathyroid hormone in circulation of man. Nature: New Biology. 1972; **238**(83):152-154

[5] Zhang C-X, Weber BV, Thammavong J, Grover TA, Wells DS. Identification of carboxyl-terminal peptide fragments of parathyroid hormone in human plasma at low-picomolar levels by mass spectrometry. Analytical Chemistry. 2006;**78**(5):1636-1643

[6] D'Amour P, Brossard JH, Räkel A, Rousseau L, Albert C, Cantor T. Evidence that the amino-terminal composition of non-(1-84) parathyroid hormone fragments starts before position 19. Clinical Chemistry. 2005;**51**(1):169-176

[7] D'Amour P, Brossard JH, Rousseau L, Nguyen-Yamamoto L, Nassif E, Lazure C, et al. Structure of non-(1-84) PTH fragments secreted by parathyroid glands in primary and secondary hyperparathyroidism. Kidney International. 2005;**68**(3):998-1007

[8] Potts JT. Parathyroid hormone: Past and present. Journal of Endocrinology. 2005;**187**(3):311-325

[9] Yamashita H, Gao P, Cantor T, Futata T, Murakami T, Uchino S, et al. Large carboxy-terminal parathyroid hormone (PTH) fragment with a relatively longer half-life than 1-84 PTH is secreted directly from the parathyroid gland in humans. European Journal of Endocrinology. 2003;**149**(4):301-306

[10] Yamashita H, Cantor T, Uchino S, Watanabe S, Ogawa T, Moriyama T, et al. Sequential changes in plasma intact and whole parathyroid hormone levels during parathyroidectomy for secondary hyperparathyroidism. World Journal of Surgery. Feb 2005;**29**(2):169-173

[11] Berson SA, Yalow RS, Aurbach GD, Potts JT. Immunoassay of bovine and human parathyroid hormone. Proceedings of the National Academy of Sciences. 1963;**49**(5):613-616

[12] D'Amour P, Labelle F, Lecavalier L, Plourde V, Harvey D. Influence of serum Ca concentration on circulating molecular forms of PTH in three species. American Journal of Physiology. Endocrinology and Metabolism. 1986;**251**(6):E680-E687

[13] Brossard JH, Cloutier M, Roy L, Lepage R, Gascon-Barré M, D'Amour P. Accumulation of a non-(1-84) molecular form of parathyroid hormone (PTH) detected by intact PTH assay in renal failure: Importance in the interpretation of PTH values. The Journal of Clinical Endocrinology and Metabolism. 1996;**81**(11):3923-3929

[14] Nussbaum SR, Zahradnik RJ, Lavigne JR, Brennan GL, Nozawa-Ung K, Kim LY, et al. Highly sensitive two-site immunoradiometric assay of parathyrin, and its clinical utility in evaluating patients with hypercalcemia. Clinical Chemistry. 1987;**33**(8):1364-1367

[15] Lepage R, Roy L, Brossard JH, Rousseau L, Dorais C, Lazure C, et al. A non-(1-84) circulating parathyroid hormone (PTH) fragment interferes significantly with intact PTH commercial assay measurements in uremic samples. Clinical Chemistry. 1998;**44**(4):805-809

[16] John MR, Goodman WG, Gao P, Cantor TL, Salusky IB, Juppner H. A novel immunoradiometric assay detects full-length human PTH but not amino-terminally truncated fragments: Implications for PTH measurements in renal failure. The Journal of Clinical Endocrinology and Metabolism. 1999;**84**(11):4287-4290

[17] Cavalier E, Vasikaran S, Bhattoa HP, Heijboer AC, Makris K, Ulmer CZ. The path to the standardization of PTH: Is this a realistic possibility? a position paper of the IFCC C-BM. Clinica Chimica Acta. 2021;**515**:44-51

[18] Gao P, Scheibel S, D'Amour P, John MR, Rao SD, Schmidt-Gayk H, et al. Development of a novel immunoradiometric assay exclusively for biologically active whole parathyroid hormone 1-84: Implications for improvement of accurate assessment of parathyroid function. Journal of Bone and Mineral Research. 2001;**16**(4):605-614

[19] Sturgeon C., Falconer H, Al-Sadie R. UK National External Quality Assessment Scheme for PTH, ACTH & hCT Annual Review 2015–2016

[20] Sturgeon CM, Sprague SM, Metcalfe W. Variation in parathyroid hormone immunoassay results - A critical governance issue in the management of chronic kidney disease. Nephrology, Dialysis, Transplantation. 2011;**26**(11):3440-3445

[21] Souberbielle JC, Boutten A, Carlier MC, Chevenne D, Coumaros G, Lawson-Body E, et al. Inter-method variability in PTH measurement: Implication for the care of CKD patients. Kidney International. 2006;**70**(2):345-350

[22] Souberbielle JC, Friedlander G, Cormier C. Practical considerations in PTH testing. Clinica Chimica Acta. 2006;**366**(1-2):81-89

[23] La'ulu SL, Roberts WL. Performance characteristics of six intact parathyroid hormone assays. American Journal of Clinical Pathology. 2010;**134**(6):930-938

[24] Cavalier E, Delanaye P, Lukas P, Carlisi A, Gadisseur R, Souberbielle JC. Standardization of DiaSorin and Roche automated third generation PTH assays with an International Standard: Impact on clinical populations. Clinical Chemistry and Laboratory Medicine. 2014;**52**(8):1137-1141

[25] Cavalier E, Salsé M, Dupuy AM, Bargnoux AS, Watar F, Souberbielle JC, et al. Establishment of reference values in a healthy population and interpretation of serum PTH concentrations in hemodialyzed patients according to the KDIGO Guidelines using the Lumipulse® G whole PTH (3rd generation) assay. Clinical Biochemistry. 2018;**54**:119-122

[26] Rubin MR, Silverberg SJ, D'Amour P, Brossard JH, Rousseau L, Sliney J, et al. An N-terminal molecular form of parathyroid hormone (PTH) distinct from hPTH(1-84) is overproduced in parathyroid carcinoma. Clinical Chemistry. 2007;**53**(8):1470-1476

[27] Räkel A, Brossard JH, Patenaude JV, Albert C, Nassif E, Cantor T, et al. Overproduction of an amino-terminal form of PTH distinct from human PTH(1-84) in a case of severe primary hyperparathyroidism:

Influence of medical treatment and surgery. Clinical Endocrinology. 2005;**62**(6):721-727

[28] Cavalier E, Betea D, Schleck ML, Gadisseur R, Vroonen L, Delanaye P, et al. The third/second generation PTH assay ratio as a marker for parathyroid carcinoma: Evaluation using an automated platform. The Journal of Clinical Endocrinology and Metabolism. 2014;**99**(3):E453-E457

[29] Hanon EA, Sturgeon CM, Lamb EJ. Sampling and storage conditions influencing the measurement of parathyroid hormone in blood samples: A systematic review. Clinical Chemistry and Laboratory Medicine. 2013;**51**:1925-1941

[30] Sturgeon CM, Sprague S, Almond A, Cavalier E, Fraser WD, Algeciras-Schimnich A, et al. Perspective and priorities for improvement of parathyroid hormone (PTH) measurement – A view from the IFCC Working Group for PTH. Clinica Chimica Acta. 2017;**467**:42-47

[31] Yan R, Lou A, Watts G, Tarr H, Smith H, Kinney L, et al. Comparison of Becton Dickinson Vacutainer rapid serum tube with the serum separator tube for routine chemistry and immunoassay tests. Journal of Clinical Pathology. 2014;**67**(7):599-604

[32] La'ulu SL, Straseski JA, Schmidt RL, Genzen JR. Thrombin-mediated degradation of parathyroid hormone in serum tubes. Clinica Chimica Acta. 2014;**437**:191-196

[33] Bolstad N, Warren DJ, Nustad K. Heterophilic antibody interference in immunometric assays. Best Practice & Research. Clinical Endocrinology & Metabolism. 2013;**27**(5):647-661

[34] Cavalier E, Delanaye P, Vranken L, Bekaert AC, Carlisi A, Chapelle JP, et al. Interpretation of serum PTH concentrations with different kits in dialysis patients according to the KDIGO guidelines: Importance of the reference (normal) values. Nephrology, Dialysis, Transplantation. 2012;**27**(5):1950-1956

[35] Touvier M, Deschasaux M, Montourcy M, Sutton A, Charnaux N, Kesse-Guyot E, et al. Interpretation of plasma PTH concentrations according to 25OHD status, gender, age, weight status, and calcium intake: Importance of the reference values. The Journal of Clinical Endocrinology and Metabolism. 2014;**99**(4):1196-1203

[36] Hocher B, Oberthür D, Slowinski T, Querfeld U, Schaefer F, Doyon A, et al. Modeling of oxidized PTH (oxPTH) and non-oxidized PTH (n-oxPTH) receptor binding and relationship of oxidized to non-oxidized PTH in children with chronic renal failure, adult patients on hemodialysis and kidney transplant recipients. Kidney & Blood Pressure Research. 2013;**37**(4-5):240-251

[37] Ursem SR, Vervloet MG, de Jongh RT, Heijboer AC. Oxidation of parathyroid hormone. Clinica Chimica Acta. 2020;**506**:84-91

[38] Daenen K, Andries A, Mekahli D, Van Schepdael A, Jouret F, Bammens B. Oxidative stress in chronic kidney disease. Pediatric Nephrology. 2019;**34**(6):975-991

[39] Hocher B, Armbruster FP, Stoeva S, Reichetzeder C, Grön HJ, Lieker I, et al. Measuring parathyroid hormone (PTH) in patients with oxidative stress - do we need a fourth generation parathyroid hormone assay? PLoS One. 2012;**7**(7):e40242

[40] Tepel M, Armbruster FP, Grön HJ, Scholze A, Reichetzeder C, Roth HJ, et al. Nonoxidized, biologically active

parathyroid hormone determines mortality in hemodialysis patients. The Journal of Clinical Endocrinology and Metabolism. 2013;**98**(12):4744-4751

[41] Seiler-Mussler S, Limbach AS, Emrich IE, Pickering JW, Roth HJ, Fliser D, et al. Association of nonoxidized parathyroid hormone with cardiovascular and kidney disease outcomes in chronic kidney disease. Clinical Journal of the American Society of Nephrology. 2018;**13**(4):569-576

[42] Barreto FC, Barreto DV, Moysés RMA, Neves KR, Canziani MEF, Draibe SA, et al. K/DOQI-recommended intact PTH levels do not prevent low-turnover bone disease in hemodialysis patients. Kidney International. 2008;**73**(6):771-777

[43] Almond A, Ellis AR, Walker SW. Current parathyroid hormone immunoassays do not adequately meet the needs of patients with chronic kidney disease. Annals of Clinical Biochemistry. 2012;**49**(1):63-67

[44] KDIGO 2017 Clinical Practice Guideline Update for the Diagnosis, Evaluation, Prevention, and Treatment of Chronic Kidney Disease–Mineral and Bone Disorder (CKD-MBD). Kidney International. Supplement. 2017;**7**(1):1-59

[45] Yalla N, Bobba G, Guo G, Stankiewicz A, Ostlund R. Parathyroid hormone reference ranges in healthy individuals classified by vitamin D status. Journal of Endocrinological Investigation. 2019;**42**(11):1353-1360

[46] Souberbielle JC, Massart C, Brailly-Tabard S, Cormier C, Cavalier E, Delanaye P, et al. Serum PTH reference values established by an automated third-generation assay in Vitamin D-replete subjects with normal renal function: Consequences of diagnosing primary hyperparathyroidism and the classification of dialysis patients. European Journal of Endocrinology. 2016;**174**(3):315-323

[47] Eastell R, Brandi ML, Costa AG, D'Amour P, Shoback DM, Thakker RV. Diagnosis of asymptomatic primary hyperparathyroidism: Proceedings of the fourth international workshop. The Journal of Clinical Endocrinology and Metabolism. 2014;**99**(10):3570-3579

[48] Fillée C, Keller T, Mourad M, Brinkmann T, Ketelslegers JM. Impact of vitamin D-related serum PTH reference values on the diagnosis of mild primary hyperparathyroidism, using bivariate calcium/PTH reference regions. Clinical Endocrinology. 2012;**76**(6):785-789

[49] Zhu CY, Sturgeon C, Yeh MW. Diagnosis and management of primary hyperparathyroidism. JAMA: The Journal of the American Medical Association. 2020;**323**(12):1186-1187

[50] Smit MA, Van Kinschot CMJ, Van Der Linden J, Van Noord C, Kos S. Clinical guidelines and PTH measurement: Does assay generation matter? Endocrine Reviews. 2019;**40**(6):1468-1480

[51] Orloff LA, Wiseman SM, Bernet VJ, Fahey TJ, Shaha AR, Shindo ML, et al. American thyroid association statement on postoperative hypoparathyroidism: Diagnosis, prevention, and management in adults. Thyroid. 2018;**28**(7):830-841

[52] Mazotas IG, Wang TS. The role and timing of parathyroid hormone determination after total thyroidectomy. Gland Surgery. 2017;**6**(Suppl. 1):S38-S48

[53] Ennis JL, Worcester EM, Coe FL, Sprague SM. Current recommended 25-hydroxyvitamin D targets for chronic kidney disease management may be too low. Journal of Nephrology. 2016;**29**(1):63-70

[54] Strugnell SA, Sprague SM, Ashfaq A, Petkovich M, Bishop CW. Rationale for raising current clinical practice guideline target for serum 25-hydroxyvitamin D in chronic kidney disease. American Journal of Nephrology. 2019;**49**(4):284-293

[55] Couchman L, Taylor DR, Krastins B, Lopez MF, Moniz CF. LC-MS candidate reference methods for the harmonisation of parathyroid hormone (PTH) measurement: A review of recent developments and future considerations. Clinical Chemistry and Laboratory Medicine. 2014;**52**:1251-1263

[56] Kumar V, Barnidge DR, Chen LS, Twentyman JM, Cradic KW, Grebe SK, et al. Quantification of serum 1-84 parathyroid hormone in patients with hyperparathyroidism by immunocapture in situ digestion liquid chromatography-tandem mass spectrometry. Clinical Chemistry. 2010;**56**(2):306-313

[57] Kritmetapak K, Losbanos LA, Hines JM, O'Grady KL, Ulmer CZ, Vesper HW, et al. Chemical characterization and quantification of circulating intact PTH and PTH fragments by high-resolution mass spectrometry in chronic renal failure. Clinical Chemistry. 2021;**67**(6):843-853

Chapter 3

TBS (Trabecular Bone Score) Used for Evaluation of Patients with Primary and Secondary Hyperparathyroidism

Lívia Marcela dos Santos and Bruno Marcos Mazoca Orozco

Abstract

The increase in parathyroid hormone (PTH) is related to a worse quality of bone mass. Bone densitometry, as an areal bone measurement, is not always able to assess bone microarchitecture. TBS (trabecular bone score) is a software that evaluates bone microarchitecture from the image of the lumbar spine obtained by bone densitometry. The articles have shown an excellent correlation of TBS with the risk of fragility fracture, especially in the individual who has primary hyperparathyroidism. We suggest that TBS may be an excellent method for assessing bone fragility in patients with hyperparathyroidism, especially if TBS is associated with bone densitometry.

Keywords: trabecular bone score, hyperparathyroidism

1. Introduction

TBS (trabecular bone score) is a parameter that indirectly measures bone microarchitecture, which is a limitation of DXA (dual-energy X-ray absorptiometry) [1, 2]. This assessment of bone microarchitecture is performed by the grayscale measurement of the lumbar spine image performed by DXA. TBS has the ability to assess the three-dimensional microarchitecture in an image obtained by two-dimensional DXA [3, 4]. The correlation that TBS presents with the parameters of computed microtomography proves its effectiveness. TBS shows a positive correlation with trabecular bone volume (BV) to tissue volume (TV) ratio (BV-to-TV), number of trabeculae, and their connectivity and stiffness; it has negative correlations with the space between trabecular and with the structure model index, measurement of rods and plates in trabecular bone [5, 6].

In a study involving 123 women and men, negative associations were found between TBS, structure model index (adjusted R2 = 69.1%), and space between trabeculae (adjusted R2 = 68.4%), positive associations were also observed between TBS and the number of trabeculae (adjusted R2 = 79.5%), and BV-to-TV (adjusted R2 = 0.830). The adjustments performed in this study were age, LS-BMD, sex, and anterior vertebral fracture (AVF) [7].

The classification of TBS is made: TBS greater than 1.310, reflects a normal bone microarchitecture, TBS between 1310 and 1230 reflects a moderately compromised bone microarchitecture, and TBS less than 1230 reflects a compromised bone microarchitecture, correlating with fragility fracture risk. In addition, it is important to remember that TBS is limited to women younger than 50 years and also with a body mass index (BMI) less than 15 and greater than 37 kg/m^2.

In 2015 the ISCD (International Society for Clinical Densitometry) officially positioned itself about TBS, with the following guidelines:

a. TBS is associated with the risk of vertebral, hip, and major osteoporotic fractures in postmenopausal women.

b. TBS can be associated with FRAX and with bone densitometry in the prediction of fractures in postmenopausal women and in elderly men [8].

2. Limitations of bone densitometry

The diagnosis of osteoporosis is currently based on bone mineral density (BMD) assessed by DXA. DXA evaluates the lumbar spine (L1-L4 vertebrae), femoral neck, total femur and radius sites. The data provided by the densitometer are in g/cm2, through these data we can compare the individual with the young adult population. The number of standard deviations coming from this evaluation generates the T-Score. The Z-Score evaluates the standard deviations comparing the individual to the population of the same age.

Through this analysis we classify the T-Score, based on the areal measurement of BMD:

- If the T-Score < −2.5 in any densitometric site is analyzed: We have a diagnosis of osteoporosis.
- A T-Score value <−2.5 is associated with a high risk of fragility fractures.

Osteopenia was classified as T-Score between 1.1 and 2.4, and the risk of fragility fracture is lower, but still exists. In addition, we have a higher absolute number of women with osteopenia compared to women with osteoporosis. Therefore, TBS is essential to distinguish women with osteopenia and increased fracture risk, because they have a degraded bone microarchitecture.

However, DXA offers some limitations, mainly because it makes a two-dimensional measurement, thus considers bone density in an areal manner, and does not have the ability to measure bone microarchitecture. Thus, when evaluating the lumbar spine image captured by DXA the TBS is able to add important information. Some studies have evaluated the ability of TBS to predict osteoporotic fractures and it has been shown to be an accurate predictor of fractures [9–12].

Thereby DXA may be an insufficient tool for determining bone strength and whether there is a satisfactory drug response for osteoporosis, mainly because it fails to measure bone microarchitecture [13–17].

TBS is a software that is easily applied in clinical practice, that is after the patient has bone densitometry the TBS can be analyzed, no additional examination is required, it is also not an invasive examination or one that will expose the patient to further radiation [4].

3. The importance of TBS (trabecular bone score) in bone fracture

With the increase in life expectancy, calcium metabolism pathologies are progressively increasing, among them the most common, osteoporosis, but also primary hyperparathyroidism (PHPT) and secondary hyperparathyroidism (SHPT). Thereby, new technologies have been emerging to improve the assessment of fragility bone fracture risk.

There is no consensus on the best tool to predict fragility fractures, but without a doubt, TBS can contribute to the early diagnosis of a patient's fracture risk. TBS can be used in association with DXA, that is, we can evaluate the T-Score of bone densitometry and the absolute value of TBS. Another way to use TBS is in association with FRAX (Fracture Risk Assessment Tool).

FRAX is a tool that calculates the risk of major and minor fractures in 10 years. The use is simple, it is a digital questionnaire, which the doctor will fill in age, sex, weight, height, if the individual has had a previous fracture, if his parents had a hip fracture, if he is a smoker, if he takes glucocorticoids chronically, if he is diagnosed with rheumatoid arthritis, if he is diagnosed with secondary osteoporosis, and if he uses alcohol more than 3 times a week. Then the doctor adds the BMD value of the patient's femoral neck. It is considered that the patient should be treated if the FRAX shows more than a 3% chance of hip fracture or more than 20% of osteoporosis-related fractures in 10 years.

Since the FRAX is a tool that estimates the risk of fragility fractures, it can also be adjusted by the TBS showing greater sensitivity to diagnose fractures. A study conducted in the province of Manitoba involving over 3000 men, with a mean age of 67 years, found that the TBS-adjusted FRAX was able to predict major osteoporotic fractures and hip fractures [18]. The FRAX is not an algorithm that can perform follow-up as it has few modifiable characteristics. However, FRAX associated with TBS can be a follow-up tool, especially for individuals who have rheumatoid arthritis and fracture risk [19]. Thus the online FRAX has an adjustment with the TBS option [20].

We can consider that the association of data such as FRAX, DXA, and TBS increases the chance of predicting vertebral and hip fracture.

4. Parathyroid hormone (PTH) and bone metabolism

Excess parathyroid hormone (PTH), a characteristic present in all forms of hyperparathyroidism, is related to decreased bone mass and increased risk of fractures.

PTH, a protein hormone secreted by the parathyroid glands, has an action on bone tissue, on the kidneys, and indirectly on the intestine. Its action in bone tissue depends mainly on PTH receptors, which have been located in osteoblasts and pluripotent progenitor cells of mesenchymal lineage, but these receptors have not been identified in osteoclasts. We know that bone remodeling is a coupled process, that is, continuous stimulation of osteoblasts by excess PTH can generate increased bone resorption.

We know that this hormone stimulation on osteoblast can indirectly activate osteoclast through osteoprotegerin (OPG) has the ability to bind to the membrane receptor (RANK) on hematopoietic progenitor cells inducing differentiation into osteoclasts, and also stimulates the production of interleukin −6 which has a role in stimulating the production and activation of osteoclast formation [21].

Primary hyperparathyroidism (PHPT) is a frequent endocrine disease consisting of hypercalcemia and elevated or inappropriately normal PTH levels. With the ease of ordering laboratory tests, especially in the last three decades, the clinical presentation of PHPT has changed from symptomatic with frequent cystic fibrous osteitis to symptomatic and asymptomatic forms of PHPT.

The classic bone clinical presentation of this syndrome is subperiosteal bone resorption, brown tumors, generalized cystic fibrous osteitis, fragility fractures, and osteopenia or osteoporosis. The bone manifestation is related to the degree of hypercalcemia presented. So, the more severe hypercalcemia, the highest degradation of bone mass. Although the most important involvement is in cortical bone, which is located in the hip, the trabecular bone is also affected, seen densitometrically by the lumbar spine [22].

A double-blind study looked at 30 patients with PHPT confirmed by surgery and pathologically confirmed with MRI or CT scans. All had osteopenia, 60% bone resorption, 40% subperiosteal resorption, and more than 30% cortical bone resorption, and subchondral resorption. There were 19 (63.3%) cases with cystic fibrous osteitis/brown tumor. There were 5 (16.7%) patients who had a pathological fracture. The skeletal disease of HPTP should be differentiated from the osteolytic metastatic bone tumor, fibrous osteodysplasia, giant cell bone tumor, and aneurysmal bone cysts [23].

5. Why should we use the trabecular bone score (TBS) as an evaluative parameter in primary hyperparathyroidism (PHPT)?

In primary hyperparathyroidism we have a decrease in bone mass, increasing the risk of vertebral and nonvertebral fractures in these individuals, and the bone mass is classically analyzed by bone densitometry. The sites analyzed in densitometry are the lumbar spine (first to fourth lumbar vertebrae), the hip (involving the femoral neck and total femur), and the distal radius.

The definitive treatment for PHPT is surgery of the affected gland, but not all patients are chosen for surgery, there are well-defined criteria. One of these criteria involves the evaluation of bone mass, that is, having a diagnosis of osteoporosis by bone densitometry. But many individuals with PHPT who have a fragility fracture only have osteopenia. Therefore, other bone mass assessment techniques can be associated with densitometry to predict fragility fractures in this population, such as TBS. Furthermore, it was evidenced that densitometry might not predict vertebral fracture risk because probably the damage in trabecular bone was more in bone microarchitecture; TBS proposes to quantify bone microarchitecture indirectly [22, 24–26].

In individuals with HPTP, the TBS is able to assess bone microarchitecture and predict fracture risk [27]. One study evaluated the significant correlation between TSB and parameters measured on HRpQCT [28].

In a study conducted in Brazil in 2021, we have the analysis of 64 individuals with primary hyperparathyroidism before parathyroidectomy, an analysis of vertebral fragility fracture was performed and TBS was able to predict this type of fracture, but

DXA could not. Also, in that study, we have that most the vertebral fractures happened in patients with osteopenia [29]. Other articles also demonstrate the ability of TBS to help predict fragility fracture in individuals with PHPT [30].

In a study that evaluated over 150 patients with PTH, with only 10% being male, means age 59 years, mean BMI 26.2 ± 4.8 kg/m^2, 89% had a diagnosis of osteoporosis/osteopenia by LS-DXA. In the patients analyzed the fracture distribution was: 7.6% with vertebral fractures, and 13.2% with non-vertebral fractures. The mean TBS was in the partially degraded range (1.258 ± 0.115); 32% of patients had degraded microarchitecture (TBS ≤ 1.20), 51% had partially degraded microarchitecture (TBS > 1.20 and <1.35) and 17% had normal TBS. TBS correlated positively with BMD in the lumbar spine and femoral neck, and negatively with age and years since menopause. Patients with vertebral fractures had mean TBS values in the degraded range, significantly lower than those without vertebral fractures. Only 9.7% of patients with degraded or partially degraded TBS microarchitecture had normal lumbar spine T-Score scores, none with vertebral fractures [27].

Obesity has been shown to be a limiting factor in the analysis of skeletal microarchitecture when assessing TBS. The influence of adiposity in the abdominal region on skeletal microstructure in primary hyperparathyroidism has not yet been well evaluated by studies. An observational study evaluated the effect of obesity on TBS and bone mineral density (BMD) in individuals with PHPT at baseline and up to 2 years after parathyroidectomy. The study participants consisted of 30 men and women with this disease, undergoing parathyroid surgery. There were notable improvements in BMD of the lumbar spine and femoral neck in the obese and non-obese individuals, but no difference in TBS values in both groups at 24 months post-parathyroidectomy. Obese individuals had more degraded TBS values compared to non-obese individuals. In this study, obesity was associated with more degraded skeletal microarchitecture as measured by TBS in PHPT, despite similar values in bone density by DXA compared to non-obese subjects. TBS values did not improve post-parathyroidectomy in obese or non-obese subjects [31].

In a cohort of symptomatic patients with HPTP, including postmenopausal, premenopausal, and male patients, we showed that TBS was in the partially degraded range but was not independently associated with fractures.

6. TBS and secondary hyperparathyroidism

Although TBS is well documented to predict osteoporotic fractures, little is known about secondary hyperparathyroidism. Low vitamin D levels are known to increase bone turnover and weaken bone architecture. In addition, 25(OH) vitamin D has a preferential action on cortical bone compared to trabecular bone.

An article was published on its correlation with 25 hydroxyvitamin D in Lebanese men. In this study, which involved both women and men, a correlation was seen between 25OH vitamin D and TBS. Fifty-four men and 61 women between the ages of 18 and 35 years were evaluated. Participants with 25(OH)D insufficiency (between 21 and 29 ng/mL) were 55.7%, and those with 25(OH)D deficiency (≤ 20 ng/mL) were 11.4%. TBS correlated positively with 25(OH)D in men ($r = 0.393$; $p < 0.05$) and women ($r = 0.324$; $p < 0.05$). In both sexes, TBS was significantly higher in 25(OH)D-sufficient participants (≥30 ng/mL). In this study vitamin D positively affects bone health and suggests that maintaining an adequate level of vitamin D may be essential for optimal TBS values [32].

Many metabolic diseases are reversible after kidney transplantation, however, this is not the case with bone disease. The fracture rate can be as high as 44% following kidney transplantation, with most fractures being vertebral. These fractures can occur due to decreased bone mass and impaired bone microarchitecture. Previous studies have correlated TBS with trabecular and cortical bone volume, width measured by bone biopsy, and histomorphometry in patients with chronic renal failure disease [33].

In patients with chronic kidney disease, there was a correlation between LS BMD and TBS in prerenal transplantation. Patients with tertiary hyperparathyroidism before transplantation had a lower TBS even after kidney transplantation. Baseline parathyroid status continued to impact TBS BMD at 6 months post-transplant with patients with tertiary hyperparathyroidism at baseline having the lowest TBS and BMD at that time. Although successful kidney transplantation should in theory address the etiopathogenesis of bone disease arising from preceding renal failure, our finding, as well as that of others, suggests that in reality hyperparathyroidism, when very advanced, may result in consequences that persist up to the year following transplantation [34, 35].

The only study that proposed a value for TBS that supposedly discriminated between subjects with and without fracture was conducted in a predominantly Caucasian population in Canada. In that study, as mentioned earlier, almost 90% of TBS measurements were done in the first year after kidney transplantation and the mean TBS in patients who did not have a fracture was 1.37 0(0.125) vs. 1.30 (0.144) in those who did [36].

7. Conclusion

The TBS is a tool that can be easily applied in clinical practice and adds information regarding fragility fractures and osteoporotic fractures in patients with primary hyperparathyroidism. It is also worth mentioning that other pathologies such as secondary and tertiary hyperparathyroidism and other bone health conditions are candidates to have TBS as a bone study tool since they also predispose to fragility and osteoporotic fractures.

Author details

Lívia Marcela dos Santos[1*] and Bruno Marcos Mazoca Orozco[2]

1 Department of Medicine, Endocrinology Unit, Federal University of – São Paulo (UNIFESP), São Paulo, SP, Brazil

2 Department of Medicine Student of the Medical School of São Camilo, São Paulo, Brazil

*Address all correspondence to: ndocrinolivia@gmail.com

IntechOpen

© 2022 The Author(s). Licensee IntechOpen. This chapter is distributed under the terms of the Creative Commons Attribution License (http://creativecommons.org/licenses/by/3.0), which permits unrestricted use, distribution, and reproduction in any medium, provided the original work is properly cited. (cc) BY

References

[1] Shevroja E, Lamy O, Kohlmeier L, Koromani F, Rivadeneira F, Hans D. Use of Trabecular Bone Score (TBS) as a complementary approach to dual-energy X-ray absorptiometry (DXA) for fracture risk assessment in clinical practice. Journal of Clinical Densitometry: The Official Journal of the International Society for Clinical Densitometry. 2017;**20**(3):334-345

[2] Ulivieri FM, Silva BC, Sardanelli F, Hans D, Bilezikian JP, Caudarella R. Utility of the Trabecular Bone Score (TBS) in secondary osteoporosis. Endocrine. 2014;**47**(2):435-448

[3] Hans D, Barthe N, Boutroy S, Pothuaud L, Winzenrieth R, Krieg M-A. Correlations between trabecular bone score, measured using anteroposterior dual-energy X-ray absorptiometry acquisition, and 3-dimensional parameters of bone microarchitecture: An Experimental Study on Human Cadaver Vertebrae. Journal of Clinical Densitometry: The Official Journal of the International Society for Clinical Densitometry. 2011;**14**(3):302-312

[4] Silva BC, Leslie WD, Resch H, Lamy O, Lesnyak O, Binkley N, et al. Trabecular bone score: A noninvasive analytical method based upon the DXA image. Journal of Bone and Mineral Research: The Official Journal of the American Society for Bone and Mineral Research. 2014;**29**(3):518-530

[5] Roux JP, Wegrzyn J, Boutroy S, Bouxsein ML, Hans D, Chapurlat R. The Predictive Value of Trabecular Bone Score (TBS) on whole lumbar vertebrae mechanics: An Ex Vivo Study. Osteoporosis International: A Journal Established as Result of Cooperation between the European Foundation for Osteoporosis and the National Osteoporosis Foundation of the USA. 2013;**24**(9):2455-2460

[6] Winzenrieth R, Michelet F, Hans D. Three-dimensional (3D) microarchitecture correlations with 2D projection image gray-level variations assessed by trabecular bone score using high-resolution computed tomographic acquisitions: Effects of resolution and noise. Journal of Clinical Densitometry: The Official Journal of the International Society for Clinical Densitometry. 2013;**16**(3):287-296

[7] Muschitz C, Kocijan R, Haschka J, Pahr D, Kaider A, Pietschmann P, et al. TBS reflects trabecular microarchitecture in premenopausal women and men with idiopathic osteoporosis and low-traumatic fractures. Bone. 2015;**79**(October):259-266

[8] Silva BC, Broy SB, Boutroy S, Schousboe JT, Shepherd JA, Leslie WD. Fracture risk prediction by non-BMD DXA measures: The 2015 ISCD official positions Part 2: Trabecular bone score. Journal of Clinical Densitometry: The Official Journal of the International Society for Clinical Densitometry. 2015;**18**(3):309-330

[9] Cummings SR, Nevitt MC, Browner WS, Stone K, Fox KM, Ensrud KE, et al. Risk factors for hip fracture in white women. The New England Journal of Medicine. 1995;**332**(12):767-773

[10] Hordon LD, Raisi M, Aaron JE, Paxton SK, Beneton M, Kanis JA. Trabecular architecture in women and men of similar bone mass with and without vertebral fracture: I. Two-dimensional histology. Bone. 2000;**27**(2):271-276

[11] McClung MR. Do current management strategies and guidelines adequately address fracture risk? Bone. 2006;**38**(2 Suppl 2):S13-S17

[12] Schuit SCE, van der Klift M, Weel AEAM, de Laet CEDH, Burger H, Seeman E, et al. Fracture incidence and association with bone mineral density in elderly men and women: The Rotterdam Study. Bone. 2004;**34**(1):195-202

[13] Bousson V, Bergot C, Sutter B, Levitz P, Cortet B, Scientific Committee of the Groupe de Recherche et d'Information sur les Ostéoporoses. Trabecular Bone Score (TBS): Available knowledge, clinical relevance, and future prospects. Osteoporosis International: A Journal Established as Result of Cooperation between the European Foundation for Osteoporosis and the National Osteoporosis Foundation of the USA. 2012;**23**(5):1489-1501

[14] Compston J. Monitoring osteoporosis treatment. Best Practice & Research. Clinical Rheumatology. 2009;**23**(6):781-788

[15] Geusens P, Vosse D, van der Linden S. Osteoporosis and vertebral fractures in ankylosing spondylitis. Current Opinion in Rheumatology. 2007;**19**(4):335

[16] Majumdar S. A review of magnetic resonance (MR) imaging of trabecular bone micro-architecture: Contribution to the prediction of biomechanical properties and fracture prevalence. Technology and Health Care: Official Journal of the European Society for Engineering and Medicine. 1998;**6**(5-6):321-327

[17] Rice JC, Cowin SC, Bowman JA. On the dependence of the elasticity and strength of cancellous bone on apparent density. Journal of Biomechanics. 1988;**21**(2):155-168

[18] Leslie WD, Aubry-Rozier B, Lix LM, Morin SN, Majumdar SR, Hans D. Spine bone texture assessed by trabecular bone score (TBS) predicts osteoporotic fractures in men: The Manitoba Bone Density Program. Bone. 2014;**67**(October):10-14

[19] Kim D, Cho S-K, Kim JY, Choi YY, Sung Y-K. Association between trabecular bone score and risk factors for fractures in Korean female patients with rheumatoid arthritis. Modern Rheumatology / the Japan Rheumatism Association. 2016;**26**(4):540-545

[20] Schacter GI, Leslie WD. DXA-based measurements in diabetes: Can they predict fracture risk? Calcified Tissue International. 2017;**100**(2):150-164

[21] Gracitelli MEC, Vidoris AAC, Luba R, Lazaretti-Castro M. Paratormônio E Osteoporose: Encontrando O Fio Da Meada. Bases Fisiológicas Para Utilização Do PTH No Tratamento Da Osteoporose. Arquivos Brasileiros de Endocrinologia E Metabologia. 2002;**46**(3):215-220

[22] Silverberg SJ, Shane E, de la Cruz L, Dempster DW, Feldman F, Seldin D, et al. Skeletal disease in primary hyperparathyroidism. Journal of Bone and Mineral Research: The Official Journal of the American Society for Bone and Mineral Research. 1989;**4**(3):283-291

[23] Qian Z-H, Rong-Jie B, Dong Y, Xin-Min L, Xiao-Guang C, Xiang G, et al. Skeletal disease in primary hyperparathyroidism: X-ray, CT and MR imaging founding. Zhonghua yi xue za zhi. 2013;**93**(1):30-33

[24] Bilezikian JP, Brandi ML, Eastell R, Silverberg SJ, Udelsman R, Marcocci C, et al. Guidelines for the management of asymptomatic primary hyperparathyroidism: Summary Statement from the Fourth International

Workshop. The Journal of Clinical Endocrinology and Metabolism. 2014;**99**(10):3561-3569

[25] Eller-Vainicher C, Battista C, Guarnieri V, Muscarella S, Palmieri S, Salcuni AS, et al. Factors associated with vertebral fracture risk in patients with primary hyperparathyroidism. European Journal of Endocrinology / European Federation of Endocrine Societies. 2014;**171**(3):399-406

[26] Kenny AM, MacGillivray DC, Pilbeam CC, Crombie HD, Raisz LG. Fracture incidence in postmenopausal women with primary hyperparathyroidism. Surgery. 1995;**118**(1):109-114

[27] Grigorie D, Coles D, Sucaliuc A. Trabecular Bone Score (tbs) has a poor discriminative power for vertebral fractures in 153 Romanian Patients with Primary Hyperparathyroidism. Acta Endocrinologica. 2018;**14**(2):208-212

[28] Cipriani C, Abraham A, Silva BC, Cusano NE, Rubin MR, McMahon DJ, et al. Skeletal changes after restoration of the euparathyroid state in patients with hypoparathyroidism and primary hyperparathyroidism. Endocrine. 2017;**55**(2):591-598

[29] Santos LMD, Ohe MN, Pallone SG, Nacaguma IO, Kunii IS, da Silva REC, et al. Trabecular bone score (TBS) in primary hyperparathyroidism (PHPT): A useful tool? Journal of Clinical Densitometry: The Official Journal of the International Society for Clinical Densitometry. October–December 2021;**24**(4):505-666. DOI: 10.1016/j.jocd.2021.04.001

[30] Muñoz-Torres M, Córdova RM, García-Martín A, Avilés-Pérez MD, Serrano RN, Andújar-Vera F, et al. Usefulness of Trabecular Bone Score (TBS) to identify bone fragility in patients with primary hyperparathyroidism. Journal of Clinical Densitometry: The Official Journal of the International Society for Clinical Densitometry. 2019;**22**(2):162-170

[31] Tay Y-KD, Cusano NE, Rubin MR, Williams J, Omeragic B, Bilezikian JP. Trabecular bone score in obese and nonobese subjects with primary hyperparathyroidism before and after parathyroidectomy. The Journal of Clinical Endocrinology and Metabolism. 2018;**103**(4):1512-1521

[32] Alwan A, al Rassy N, Berro A-J, Rizkallah M, Matta J, Frenn F, et al. Vitamin D and trabecular bone score in a group of young lebanese adults. Journal of Clinical Densitometry: The Official Journal of the International Society for Clinical Densitometry. 2018;**21**(3):453-458

[33] Ramalho J, Marques IDB, Hans D, Dempster D, Zhou H, Patel P, et al. the trabecular bone score: Relationships with trabecular and cortical microarchitecture measured by HR-pQCT and histomorphometry in patients with chronic kidney disease. Bone. 2018;**116**(November):215-220

[34] Bouquegneau A, Salam S, Delanaye P, Eastell R, Khwaja A. Bone disease after kidney transplantation. Clinical Journal of the American Society of Nephrology: CJASN. 2016;**11**(7):1282-1296

[35] Chandran M, Hao Y, Kwee AK, Swee DS, Ng DCE, Kee TYS, et al. Addressing bone quality and bone density after renal transplantation: A prospective evaluation of the evolution of trabecular bone score and bone mineral density over the first 5 years

following renal transplantation in Asian patients. Clinical Transplantation. 2019;**33**(10):e13671

[36] Naylor KL, Garg AX, Hodsman AB, Rush DN, Leslie WD. Long-term changes in bone mineral density in kidney transplant recipients. Transplantation. 2014;**98**(12):1279-1285

Chapter 4

Perspective Chapter: Parathyroid Glands in Dentistry

Antigoni Delantoni

Abstract

Parathyroid glands are found adjacent to the thyroid, produce parathyroid hormone, and regulate with their function serum calcium levels. Parathyroid conditions are encountered in dentistry as in other health sciences. In dentistry, however, since they are not so frequently observed or encountered they are often not diagnosed, or not identified. Many patients, though they have a known history of parathyroid gland disorders or conditions, often neglect to mention it to the dentist, or in many cases the dentist omits it in the medical history section, and since in most cases they are often unaware of the dental findings, the conditions and malfunction of the glands may clinically present at patients. The dentists and medical practitioners should note in those patients, in cases of hyperparathyroidism possible brown tumors of the jaws, and in cases of hypothyroidism short stature, developmental disturbances, teeth development disturbances, etc. In this chapter, the major findings of the disorders of parathyroid glands are noted and reported as well as thoroughly described with clinical and dental radiographic findings of the conditions.

Keywords: dentistry, parathyroid glands, hyperparathyroidism, hypoparathyroidism, brown tumor

1. Introduction

Parathyroid hormone is produced by the parathyroid glands and has a key role in bone formation. These glands control the amount of calcium in our blood. Everyone has four parathyroid glands, usually located right around and behind the thyroid gland at the base of the neck. They are about the size of a grain of rice. These small glands produce parathyroid hormone, regulating calcium levels [1].

2. Hyperparathyroidism

The percentage of developing some type of parathyroid gland tumor in adult people is about 1%. In most cases, the tumor or the condition will cause "hyperparathyroidism" [2]. The condition is aggressive and causes high blood calcium levels, which can lead to serious health problems. It presents when the parathyroid glands create too much parathyroid hormone in the bloodstream. Hyperparathyroidism (HPT) may be primary, secondary, and tertiary type and is featured as overproduction of parathyroid hormone.

IntechOpen

Primary hyperparathyroidism [2] may be cured by removing the adenoma or overactive parathyroid glands [3, 4]. In those without symptoms, mildly increased blood calcium levels, normal kidneys, and normal bone density monitoring may be all that is required. The deficiency of vitamin D with low serum levels should be corrected. The case of primary hyperparathyroidism is the most common type. In the developed world, the number of people affected is estimated, between one and four per thousand people. The condition occurs three times more often in women than men and is typically diagnosed between the ages of 50 and 60. Radiation exposure increases the risk of primary hyperparathyroidism. Several genetic conditions including multiple endocrine neoplasia syndromes also increase the risk [4].

In other cases of secondary hyperparathyroidism, what usually happens is the following: In patients with chronic kidney disease (CKD), impaired renal function leads to decreased vitamin D levels, which in turn causes an increase in parathyroid hormone production and contributes to the development of secondary hyperparathyroidism. The low levels of vitamin D lead to reduced calcium absorption by the intestine thus resulting in hypocalcemia and increased parathyroid hormone level secretion. This in turn leads to an increase in bone resorption. When the bone condition is caused by kidney failure in the case of secondary hyperparathyroidism, the pathology is termed renal osteodystrophy [5, 6].

Long-term secondary hyperparathyroidism, which eventually leads to hyperplasia of the parathyroid glands, can lead in a number of cases to tertiary hyperparathyroidism, which leads to a loss of response to serum calcium levels [5, 6].

In dentistry, the most frequent presentation of hyperparathyroidism is a brown tumor [5–8]. The brown tumor is a bone lesion that arises in settings of excess osteoclastic activity, such as hyperparathyroidism. The brown tumor (BT) or osteitis fibrosa cystica is a benign osseous lesion, which may result in any form of uncontrolled parathyroid hormone hypersecretion [7]. These non-neoplastic lesions present late in the untreated disease have been severely decreased by early diagnosis and successful management of hyperparathyroidism in developed countries [7, 8].

Diagnosis of brown tumors is merely presumptive, and in most cases in the jaws, it presents as an incidental finding. Histology sets the final diagnosis. Laboratory data and radiographs may be additionally used for definitive clinical diagnosis. In a radiographical examination, there are findings from the bones that can even aid the diagnosis of the condition in a large number of patients.

Brown tumor may show no changes or a generalized osteoporosis image, making it difficult to diagnose. The lesions present as sharply defined, round, or oval radiolucent areas, which may appear as multilocular (**Figure 1**). It is useful to know the serum levels of calcium and parathyroid hormone to set the differential diagnosis from giant cell lesions [8].

In hyperparathyroidism, serum parathyroid hormone level can be indicative of the diagnosis of the brown tumor, when compared with other giant cell tumors. In a rare case published by our clinic, the patient's panoramic radiography that was taken 2 years ago (**Figure 2**) showed a very early lesion with no well-defined radiolucent osteolytic lesions near the middle section of the teeth roots. In the later panoramic radiograph of the patient, the lesion had progressed to a well-defined lucency adjacent to the tooth's roots extending to the surrounding bone. The lesion was removed regionally, and the systemic management revolved primarily in reducing circulating endocrine hormone (parathyroid hormone) [9].

What should be noted in this case is that in the first panoramic radiograph though, there was some indication that the diagnosis could not have been set from

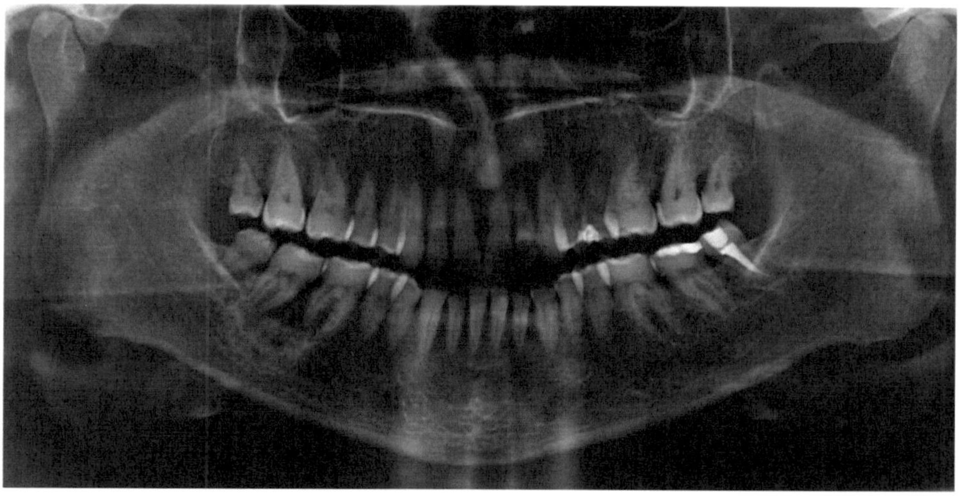

Figure 1.
A panoramic radiograph of the patient indicating a localized well-defined radioluscent area in the lower right molar region.

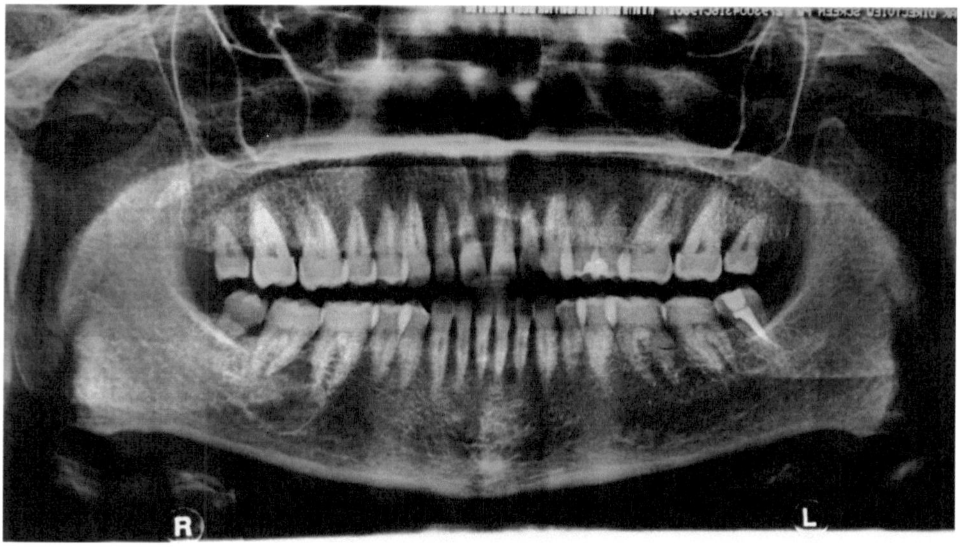

Figure 2.
Panoramic radiograph of the same patient that was taken 2 years prior to the diagnosis and hardly indicates any pathology in the area.

the radiographic image. Similar cases have been dealt accordingly with no need for further medical management. Parathyroidectomy is the treatment of choice. In our case, the brown tumor of the mandible was the first sign of the condition (hyperparathyroidism) due to inbalance of osteoblastic and osteoclastic activities, formed by increased parathyroid hormone levels and calcium phosphorous serum level regulation. Upon re-examining the patient, there was no recurrence of the lesion or presence of other similar lesions [9].

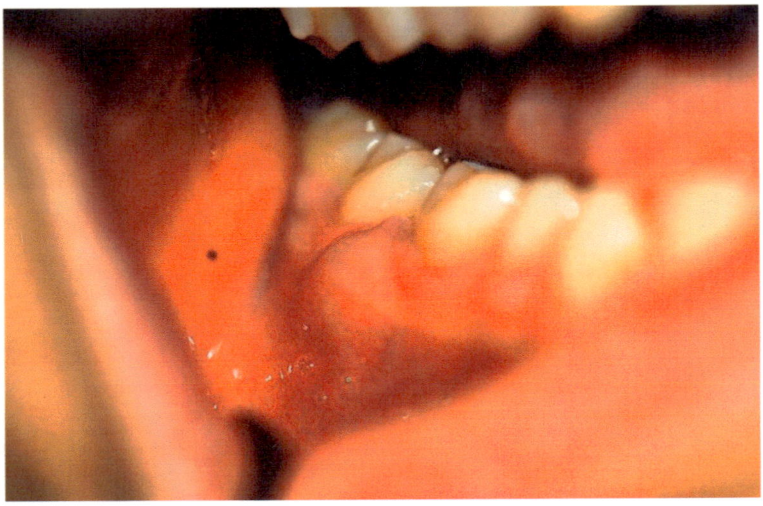

Figure 3.
Clinical presentation of the condition localized in the gingival area of the lower right molar.

Our case had the unusual primary detection of the condition in the mandible.

Clinically, there was mild swelling in the region and hardly any pain upon palpation of the lesion (**Figure 3**). It is a rare case but should alert radiologists, and one should be careful of the medical history of a patient even upon taking a plain radiograph. The differential diagnosis should be set, only after a thorough study of the patient's both medical and dental history. Dentists should be aware of the medical conditions that may cause dental problems and should be more alert to certain patients.

In conclusion, the development of brown tumors in the jaws is not one of the common findings of the disease [10]. At the clinical level, the brown tumor initially appears as a hard bone-like tumor, which shows very slow growth. Later, when the tumor erodes the cortical bone marrow, a change in its texture is observed. At a radiological level, in classical radiographs the brown tumors are depicted as single-chambered or multi-chambered, usually mixed (shading and clarifying) lesions [11–13].

3. Hypoparathyroidism

Besides hyperparathyroidism that is the most frequently produced parathyroid gland malfunction, there is a rare and more misdiagnosed case of hypoparathyroidism.

Hypoparathyroidism is decreased function of the parathyroid glands, with underproduction of parathyroid hormone. This can lead to low levels of calcium in the blood, often causing cramps and twitching of muscles or even in more advanced cases together with other symptoms. Hypoparathyroidism is a very rare condition that can be inherited, but can also present itself after thyroid or parathyroid gland surgery, and can be caused by immune system-related damage as well as several rarer causes [14–16].

Hypoparathyroidism does not represent one condition but rather a group of diseases characterized by hypocalcemia and hyperphosphatemia clinically, Basically, the result and clinical measurements set the diagnosis. These disorders that can present as hypoparathyroidism can be the result of either reduced or damaged secretion of parathyroid hormone by the parathyroid gland or failure of the target organs to respond to the hormone properly. In the last described case where the cause is not the gland itself, the condition is known as pseudohypoparathyroidism (PHP) and is associated with an elevated plasma concentration of parathyroid hormone (not originating from glandular malfunction) [14]. Fuller Albright described the first patients with this disorder in 1942, and the condition took his name and is described as Albright condition. The patients with the condition demonstrate phenotypical features that include a round face, short stature, obesity, brachydactyly, and ectopic calcifications. The features of the condition are typically found in patients with type Ia PHP, also known as Albright hereditary osteodystrophy [14–17].

Distinction between type I and type II PHP is based on the response of cyclic adenosine monophosphate (cAMP) to parathyroid hormone, which is abnormal in type I and normal in type II. These metabolic abnormalities are very rare, with a reported prevalence of only 0.79 per 100,000 individuals [18].

The major features and classical to the condition findings include a high parathormone serum concentration, high serum phosphate, and low serum calcium concentrations, all caused by the existing end-organ resistance.

Patients with PHP may or may not have the clinical features of type I of the disease, such as short stature, rounded face, and central obesity [19, 20] though in most cases they present with symptoms indicative of electrolyte disturbances (Ca and P levels). Imaging features mainly reflect abnormalities in the musculoskeletal and central nervous systems. The lateral cephalogram of the patients demonstrates the hypoplasia of the jaws (**Figure 4**). Computed tomography of the central nervous system in the condition usually reveals calcifications in the basal ganglia, specifically in the striatopallidal nuclei. Magnetic resonance imaging as it better displays the soft tissues [19], it may represent an even more sensitive modality helping in the early detection of clinical and radiographic changes in the basal ganglia.

PHP is a condition limited to only a few cases reported in the dental literature since it is a very rare condition [21].

In a rare case in our clinic, in the panoramic radiograph of the patient (**Figure 5**), there were multiple impactions of teeth with malformation of their development with mostly short roots and unusual shape. In the cone beam-computed tomography (CBCT) that was taken, the findings of PHP were thoroughly studied. Besides, the multiple impactions and the shape of the teeth calcifications of the basal ganglia were observed as an incidental finding [22]. Given the rarity of the disease and the limited literature, radiologists and dentists should be aware of the value of CBCT in the identification of incidental findings such as calcifications in the basal ganglia, which should always be reported. Only few case reports articles are present in the literature under the query "PHP and teeth," and of those, only one paper was based on the presentation of four cases, and it is the one that provides a thorough description of the dental findings of the condition [23–27]. The initial description of the teeth deformities and malformations in patients with PHP occurred in 1952 by Mackler et al., who noted numerous unerupted teeth in medially PHP patients [28]. Although in the medical literature there are numerous studies of the condition including a large number of cases of PHP, there are very few references to the associated changes that occur in the dental tissues and facial development, since the jaws are often neglected in the clinical examination of the patients and also because in most cases, there are more serious issues to be addressed and prioritized. We also have to take into consideration that with the recent advances in medicine, PHP is

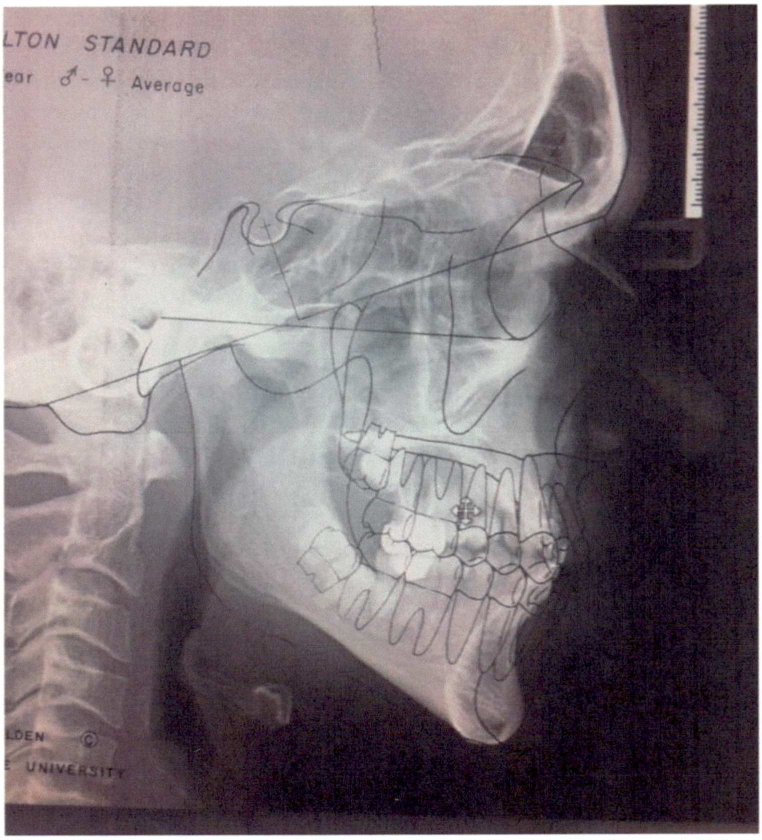

Figure 4.
Cephalometric radiograph of a patient with PHT where we can measure a smaller jaw size and reduced skeletal development.

being diagnosed early in larger numbers of patients, even before reaching adulthood and completing their skeletal growth; however, only three case reports have been published since the year 2000.

The condition though it is very rare it should be taken into account when there is a known medical condition or symptoms present and not be excluded from the differential diagnosis of the patient's pathosis [29].

In conclusion, dental findings of the condition include short roots with blunt ends, and small crowns, thin enamel, and large pulp chambers, many impacted or not properly erupted teeth, early teeth loss due to caries, and short and wide jaws with hypoplasia of both upper and lower jaw as well as the entire facial skeleton. When a number of those are present, the dentist should be alert to the possibility of the condition [23–30]. Though the diagnosis of PHP is based on laboratory findings of the patient, the dental findings should not be neglected.

4. Conclusions

Parathyroid glands are often causing various hormonal disorders and problems. The dentists should be fully alert to the conditions caused by parathyroid disorders

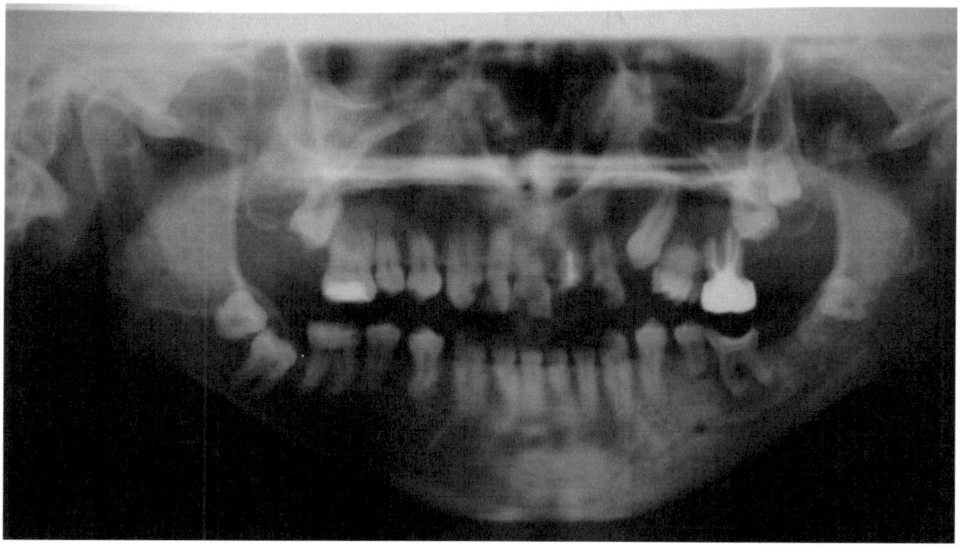

Figure 5.
The panoramic radiograph of the same patient where multiple dental impactions can be seen together with alterations to the tooth shape and size.

and bear in mind the medical history of the patients. Besides the-medical history of the patients, a detailed and thorough clinical examination may reveal pathologies that involve parathyroid hormone disorders. The evaluation of clinical and radiological pathological findings by the dentist can lead to the diagnosis of systemic diseases. The dentist may even set the first diagnosis of the systemic disease.

Conflict of interest

The authors declare no conflict of interest.

Author details

Antigoni Delantoni
Faculty of Dentistry, Department of Dentoalveolar Surgery, Implant Surgery and Radiology, School of Health Sciences, Aristotle University of Thessaloniki, Thesssaloniki, Greece

*Address all correspondence to: andelant@dent.auth.gr

IntechOpen

© 2022 The Author(s). Licensee IntechOpen. This chapter is distributed under the terms of the Creative Commons Attribution License (http://creativecommons.org/licenses/by/3.0), which permits unrestricted use, distribution, and reproduction in any medium, provided the original work is properly cited.

References

[1] Delantoni A, Lyroudia K, Rafailidis V, Chryssogonidis I. Pseudohypoparathyroidism diagnosed in adulthood: maxillofacial clinical and radiographic findings. Oral Radiology. 2017;**33**(2):153-156

[2] Minisola S, Gianotti L, Bhadada S, Silverberg SJ. Classical complications of primary hyperparathyroidism. Best Practice & Research. Clinical Endocrinology & Metabolism. 2018;**32**(6):791-803

[3] Shetty AD, Namitha J, James L. Brown tumor of mandible in association with primary hyperparathyroidism: A case report. Journal of International Oral Health. 2015;**7**(2):50-52

[4] Resendiz-Colosia JA, Rodriguez-Cuevas SA, Flores-Diaz R, et al. Evolution of maxillofacial brown tumors after parathyroidectomy in primary hyperparathyroidism. Head and Neck. 2008;**30**(11):1497-1504

[5] Fang Y, Hong G, Lu H, Guo Y, Yu W, Liu X. Brown tumor: Clinical, pathological and imaging manifestations. Zhonghua Yi Xue Za Zhi. 2015;**95**(45):3691-3694

[6] Hu J, He S, Yang J, Ye C, Yang X, Xiao J. Management of browntumor of spine with primary hyperparathyroidism: A case report and literature review. Medicine (Baltimore). 2019;**98**(14)

[7] Nassar GM, Ayus JC. Images in clinical medicine. Brown tumor in end-stage renal disease. The New England Journal of Medicine. 1999;**341**(22):1652

[8] Keyser JS, Postma GN. Brown tumor of the mandible. American Journal of Otolaryngology. 1996;**17**(6):407-410

[9] Delantoni A, Önder M, Orhan K. Brown tumor of the mandible as an initial presentation of hyperparathyroidism. European Annals of Dental Sciences. 2021;**48**(2):74-77

[10] Olsen J, Sealey C. Brown tumour of the mandible in primary hyperparathyroidism; a case report. The New Zealand Dental Journal. 2015;**111**(3):116-118

[11] Huang R, Zhuang R, Liu Y, Li T, Huang J. Unusual presentation of primary hyperparathyroidism: report of three cases. BMC Medical Imaging. 2015;**16**(15):23

[12] Gulati D, Bansal V, Dubey P, Pandey S, Agrawal A. Central giant cell granuloma of posterior maxilla: First expression of primary hyperparathyroidism. Case Reports in Endocrinology. 2015:170412. Published online 2015 Jan 26. DOI: 10.1155/2015/170412

[13] Di Daniele N, Condo S, Ferrannini M, et al. Brown tumour in a patient with secondary hyperparathyroidism resistant to medical therapy: Case report on successful treatment after subtotal parathyroidectomy. International Journal of Endocrinology. 2009: 827652. Published online 2009 Apr 16. DOI: 10.1155/2009/827652

[14] Nelson W. Textbook of Pediatrics. 10th ed. Philadelphia: W.B. Saunders Co.; 1975

[15] Levine MA. An update on the clinical and molecular characteristics of pseudohypoparathyroidism. Current Opinion in Endocrinology, Diabetes, and Obesity. 2012;**19**(6):443-451

[16] Albright F, Buvnett CH, Smith PH, Parson W. Pseudohypoparathyroidism, an example of Seabright-bantam disorder. Endocrinology. 1942;**30**:922

[17] Maeda SS, Fortes EM, Oliveira UM, Borba VC, Lazaretti-Castro M. Hypoparathyroidism and pseudohypoparathyroidism. Arq Bras Endocrinol Metabol. 2006;**50**(4):664-673

[18] Balachandar V, Pahuja J, Maddaiah VT, Collipp J. Pseudohypoparathyroidism with normal serum calcium level. American Journal of Diseases of Children. 1975;**129**(9):1092-1095

[19] Mantovani G. Clinical review: Pseudohypoparathyroidism: Diagnosis and treatment. The Journal of Clinical Endocrinology and Metabolism. 2011;**96**(10):3020-3030

[20] Monn E, Osnes JB, Oye I, Wefring KW. Pseudohypoparathyroidism: A difficult diagnosis in early childhood. Acta Paediatrica Scandinavica. 1976;**65**(4):487-493

[21] Stanbury JB, Wyngaarden JB, Fredrickson DS. The Metabolic Basis of Inherited Diseases. New York: McGraw Hill Book Company; 1966

[22] Dosi RV, Ambaliya AP, Joshi HK, Patell RD. Pseudohypoparathyroidism, rare cause of hypocalcaemia. Journal of Clinical & Diagnostic Research. 2013;7(10):2288-2289

[23] BoswellSB PDB, White EA, Gottsegen CJ, Forrester DM, Masih S, Matcuk GR Jr. Musculoskeletal manifestations of endocrine disorders. Clinical Imaging. 2014;**38**(4):384-396

[24] Kalita J, Kumar N, Maurya PK, Misra UK. A study of computed tomography scan and magnetic resonance imaging findings in pseudohypoparathyroidism. Annals of Indian Academy of Neurology. 2010;**13**(3):204-206

[25] Ritchiem GMacL. Dental manifestations of pseudohy poparathyroidism. Archives of Disease in Childhood. 1965;**40**:565-572

[26] Mackler H, Fouts JR, Birsner JW. Familiar pseudohypoparathyroidism: A report of two cases. California Medicine. 1952;77:332-336

[27] Koklu E, Kurtoglu S. Natal teeth and neonatal transient pseudohypoparathyroidism in anewborn. Journal of Pediatric Endocrinology and Metabolism. 2007;**20**(9):971

[28] Mackler H, Fouts JR, Birsner JW. Familial pseudohypoparathyroidism: A report of two cases. California Medicine. 1952;77(5):332-334

[29] Goswami M, Verma M, Singh A, Grewal H, Kumar G. Albright hereditary osteodystrophy: A rare case report. Journal of the Indian Society of Pedodontics and Preventive Dentistry. 2009;**27**(3):184-188

[30] Velez I, Bond M, Ellen S, Ede-Nichols D, Larumbe J, Oramas V, et al. Hereditaryosteodystrophy with multiple hormone resistance--a case report. The Journal of Clinical Pediatric Dentistry. 2009;**34**(1):67-69

Chapter 5

Genetic Disorders of Calcium and Phosphorus Metabolism Related with Parathyroid Glands

Ayça Dilruba Aslanger

Abstract

Calcium (Ca), phosphorus (phosphate, HPO4), and magnesium (Mg) are essential nutrients that are critical for the structural integrity and functions of the body. Therefore, disorders of calcium and phosphorus metabolism lead to serious and even life-threatening consequences such as skeletal and cardiovascular morbidity. Parathyroid hormone (PTH), calcitonin, and the active form of vitamin D (calcitriol, 1,25-dihydroxyvitamin D3) hormones are the main hormones that are responsible for regulating the calcium and phosphorus level in the blood. Hypoparathyroidism is due to insufficient circulating parathyroid hormone levels characterized by hypocalcemia and hyperphosphatemia. Besides being an isolated condition or a component of a complex syndrome, the causes of hypoparathyroidism are rarely genetic. Primary hyperparathyroidism is a disorder that results in excessive, uncontrolled production of parathyroid hormone. Rarely, primary hyperparathyroidism caused by genetic disorders is associated with an inherited familial germline mutation syndrome such as familial isolated hyperparathyroidism and multiple endocrine neoplasia type 1 and type 2A. Although genetic disorders are not the most common cause of hyper/hypoparathyroidism, molecular analyses have identified an increasing number of genes that cause loss or gain of function of genes related to calcium and phosphorus metabolism.

Keywords: parathyroid hormone; genetic endocrine conditions, hypoparathyroidism, hyperparathyroidism, pseudohypoparathyroidism

1. Introduction

Around 99% of total body calcium is present in the bones with the remaining 1% being found in the extracellular fluid and cellular organelles. Approximately 50% of total serum calcium is bound to plasma proteins, mostly albumin and globulin, and 5% is complex with citrate, lactate, and bicarbonate. The other 45% of serum calcium is an ionized and biologically active form of calcium. Serum total and ionized calcium concentrations are associated with levels of albumin, creatinine, parathyroid hormone (PTH), phosphate, and serum pH. Ca^{2+} serum concentration is regulated by a combined system including the extracellular calcium-sensing receptor (CaSR); PTH and its receptor (PTH/PTH-related protein PTHrP-1R);

calcitonin and its receptor; and vitamin D hormone system that acts on the intestinal tract, bone, and kidney. The absorption of calcium is enhanced in the intestinal tract, renal tubule, and bones in response to calcitriol (1,25OH2D3) and/or PTH. PTH and calcitriol mobilize calcium from the hydroxyapatite apatite crystal, and calcitonin secreted by the parafollicular C cells of the thyroid gland suppresses PTH secretion. PTH increases concentrations of plasma Ca^{2+}, lowers serum phosphate values, accelerates the synthesis of calcitriol, and stimulates both anabolic and catabolic effects on bone [1, 2].

The parathyroid glands develop together with the thymus from the third and fourth pharyngeal pouch. Hypoparathyroidism, also called PTH deficiency, causes hypocalcemia, hyperphosphatemia, and increased neuromuscular irritability [3]. It is usually caused by the postoperative complications of head and neck surgery in an individual. The other most common cause of hypoparathyroidism is autoimmune destruction of parathyroid tissue associated with autoimmune disorders that cause destruction or damage to the glands. In infancy, hypoparathyroidism due to delayed developmental maturation of the parathyroid is usually transient and can be resolved within the first few weeks [4, 5]. When hypoparathyroidism is persistent, it may result; from an error in embryogenesis or destruction of the parathyroid glands, disorders resulting in impaired PTH synthesis or secretion, or peripheral unresponsiveness to PTH.

2. Disorders of parathyroid gland formation

2.1 22q11 deletion syndrome

22q11.2 deletion syndrome, also known as Di George syndrome, is the most common human contiguous gene deletion caused by a microdeletion of chromosome 22 with a wide phenotypic spectrum that is highly variable even within the family. It is characterized by particularly conotruncal cardiac defects, immune deficiency, palatal anomalies, craniofacial dysmorphism, and hypoparathyroidism. Additionally, patients with 22q 11.2 deletion may have feeding difficulties, poor growth, intellectual disability, and psychiatric problems [6]. Most patients with 22q11 deletion have heterozygous 2.5–3 Mb deletions including TBX1 gene, which encodes the T-box transcription factor gene 1. Patients with point mutations in TBX1 manifest many of the clinical findings of cases with 22q11 deletions [7]. This microdeletion syndrome usually occurs de novo but can also be inherited from a parent (5–10%). The estimate that DiGeorge syndrome affects one in every 2000–4000 live births may be an underestimate as it is based on major birth defects. Because some people with deletions will have few symptoms, these cases may go undiagnosed. Hypoparathyroidism in patients with 22q11.2 deletion results from aplasia or hypoplasia of parathyroid glands. It has been reported that 30–60% of patients with hypoparathyroidism 22q11.2 causing hypocalcemia are present especially in the neonatal period [8]. Many forms of hypoparathyroidism including late-onset, transient, latent, and recurrent types have been reported in patients with 22q11 deletion syndrome [9]. Although calcium homeostasis generally improves with age, hypocalcemia tends to occur in adolescence, pregnancy, and later in life can occur during times of stress (for example, surgery or serious illness [10, 11]). It has been reported that 80% of adults with 22q11.2 deletion patients may have hypocalcemia, hypoparathyroidism, hypothyroidism, and hypomagnesemia [12]. Typical features of 22q11 deletion include thymic aplasia with impaired

T-cell mediated immunity, conotruncal cardiac defects, cleft palate, and dysmorphic facies with mid-face hypoplasia and tubular nose [13].

2.2 CHARGE syndrome – CHD7, SEMA3E

CHARGE syndrome is an acronym for Coloboma, Heart disease, Atresia of the choanae, Retarded growth and mental development, Genital anomalies, Ear malformations, and hearing loss with an incidence of 1 in 8,500–10,000 [14]. Hypoparathyroidism due to parathyroid hypoplasia may accompany as a component of CHARGE syndrome. Most patients have a heterozygous loss-of-function mutation within the coding region of the Chromodomain helicase DNA binding-7 (CHD7) gene on chromosome 8q1 [15]. CHD7 is a member of the Chromodomain helicase DNA-binding protein (CHD) family of ATP-dependent chromatin remodelers, which catalyze nucleosome movement on DNA. CHD7 functions as a positive regulator of ribosomal RNA (rRNA) biogenesis in the nucleolus and uses the energy of ATP to remodel nucleosomes. CHARGE is an autosomal dominant disease often caused by a de novo pathogenic variant. Rarely, the patients with CHARGE Syndrome inherit the mutation from an affected parent. A less common cause is due to abnormalities involving semaphorin 3E (SEMA3E) that are involved in embryonic development for neural guidance [16].

2.3 Barakat syndrome (Hypoparathyroidism, Deafness and Renal dysplasia syndrome HDRS)

Hypoparathyroidism, sensorineural deafness, and renal dysplasia syndrome (HDRS), also known as Barakat syndrome, is an autosomal dominant disorder caused by a monoallelic mutation in the GATA3 gene on chromosome 10p14 [17, 18]. The GATA3 protein is one of the transcription factors required for the development of the pharyngeal sacs and the differentiation and organization of the parathyroid glands. It is expressed in parathyroid glands as well as in the thymus, kidney, inner ear, and central nervous system. GATA3 interacts with two known transcriptional regulators of parathyroid development, GCM2 (isolated parathyroid aplasia) and MAFB, and synergistically stimulates the promoter of the PTH gene, thereby activating PTH gene transcription and regulating PTH gene expression. Barakat syndrome is characterized by the classic triad of hypoparathyroidism, sensorineural deafness, and/or renal disease. Renal manifestations may include renal dysplasia, progressive renal failure, proteinuria, glomerulonephritis, and renal agenesis. Renal abnormalities are observed in 60% of patients and are highly variable, with only a small proportion (9%) of cases progressing to end-stage renal disease [19]. Additional clinical features include congenital heart disease, hypogonadotropic hypogonadism, polycystic ovaries, retinitis pigmentosa, and intellectual disability. There is wide phenotypic variability with hypoparathyroidism ranging from asymptomatic or transient neonatal hypocalcemia to severe symptomatic or persistent hypocalcemia. Sensorineural hearing loss is typically discovered during infancy or childhood is usually bilateral and present in more than 95% of cases [20].

2.4 Sanjad-Sakati Syndrome (Hypoparathyroidism-Retardation-Dysmorphism syndrome HRDS)

The Sanjad-Sakati syndrome of congenital hypoparathyroidism, mental retardation, and dysmorphism (HRD, OMIM 241410) is an autosomal recessive disorder caused by biallelic loss-of-function mutations in TBCE (tubulin-specific chaperone E) gene on chromosome 1q42 [21]. Facial dysmorphic features include a prominent forehead, deep-set eyes, microphthalmia, depressed nasal bridge, beak nose, long philtrum, thin upper lip, micrognathia, bifid uvula, and ear abnormalities. Most patients have intrauterine and postnatal growth retardation, developmental delay, hypocalcemic seizures [22]. TBCE gene encodes a chaperone protein required for formation, folding, and stability of alpha-tubulin subunits and the formation of alpha-beta-tubulin heterodimers. Kenny-Caffey syndrome (OMIM 244460), allelic to Sanjad-Sakati syndrome, manifests cortical thickening and medullary stenosis of the long bones, osteosclerosis of the skull, and recurrent bacterial infections [23].

2.5 Mitochondrial disorders associated with hypoparathyroidism

Hypoparathyroidism may accompany mitochondrial disorders whose typical clinical features are lactic acidosis, cataracts, sensorineural deafness, and myopathy/ophthalmoplegia. The mitochondrial disorders associated with hypoparathyroidism include Kearns-Sayre syndrome, mitochondrial myopathy, encephalopathy, lactic acidosis and stroke-like episodes (MELAS), long-chain 3-hydroxyacyl-CoA dehydrogenase (LCAHD), medium-chain acyl-CoA dehydrogenase deficiency (MCADD), and combined mitochondrial trifunctional protein (MTP) [24].

3. Disorders of parathyroid gland destruction

Autoimmune polyendocrine syndrome 1 (APS1) is an autosomal recessive disorder characterized by the classic triad of autoimmune polyendocrinopathy, mucocutaneous candidiasis, and ectodermal dystrophy (APECED). There are many additional autoimmune endocrinopathies including hypoparathyroidism, hypothyroidism, insulin-dependent diabetes mellitus, primary adrenal insufficiency, and ovarian/testicular failure. Less often non-endocrine signs are pernicious anemia, hepatitis, keratoconjunctivitis, vitiligo, alopecia, malabsorption, and metaphyseal dysplasia. APS1 is caused by homozygous or compound heterozygous loss-of-function variants of the autoimmune regulator gene (AIRE, OMIM 607358) [25]. Almost all women with APECED and 80% of affected men develop hypoparathyroidism. Despite wide variability in clinical expression, there is no significant association between genotype and phenotype. The diagnosis of APECED usually requires the presence of two of the three main findings (mucocutaneous candidiasis, hypoparathyroidism, and hypoadrenocorticism), but sometimes just one of the three findings alone may be the sole manifestation of an inactivating mutation in AIRE. Later manifestations of APECED involve esophageal and oral squamous cell carcinoma, asplenia, and interstitial nephritis [26].

4. Disorders resulting in impaired PTH synthesis or secretion

The parathyroid glands develop together with the thymus from the third and fourth pharyngeal sacs under the impact of multiple transcription factors including T-box 1 (encoded by TBX1), GATA-binding factor/protein 3 (encoded by GATA3), glial cell missing, drosophila, homolog 2 (encoded by GCM2), and V-MAF musculoaponeurotic fibrosarcoma oncogene family, protein B (encoded by MAFB), an essential factor that supports in the separation of parathyroid glands from thymus and migration to the thyroid. Variants of GATA3 are associated with the syndrome of hypoparathyroidism, sensorineural deafness, and renal disease, whereas mutations in GCM2 result in both familial hypo- and hyperparathyroidism [27, 28]. Familial isolated congenital hypoparathyroidism can be transmitted as an autosomal dominant, autosomal recessive, or X-linked recessive trait resulting from loss-of-function mutations in genes necessary for differentiation of the parathyroid glands causing congenital aplasia or hypoplasia of these structures. Familial isolated hypoparathyroidism is caused by heterozygous, homozygous, or compound heterozygous mutation in the parathyroid hormone gene PTH (Familial isolated congenital hypoparathyroidism 1) gene, GCM2 gene (Familial isolated congenital hypoparathyroidism 2) [29, 30].

4.1 PTH gene

Familial isolated hypoparathyroidism-1 (FIH1) is caused by heterozygous, homozygous, or compound heterozygous mutation in the parathyroid hormone gene (PTH) on chromosome 11p15 [31]. PTH gene is composed of three exons, where its second and third exons of PTH encode the prepro-PTH sequence of 115 amino acids that is processed to intact 84 amino acids PTH that is released from the parathyroid glands in response to decreasing serum concentrations of Ca^{2+}, detected by the CaSR expressed on the plasma membrane of the parathyroid gland chief cell. Familial isolated hypoparathyroidism 1 is caused by monoallelic/biallelic mutations in the PTH gene located on chromosome 11p15.3 that impair synthesis of PTH. Since the PTH gene was discovered, only eight pathogenic variants in PTH gene have been identified. A few patients with FIH1 have been associated with heterozygous Cys18Arg mutation in PTH gene that mutation induces endoplasmic reticulum stress and subsequent apoptosis in parathyroid cells. A donor splice site mutation at the nucleotide 1 of intron 2 of PTH leads to exon 2 skipping, with loss of the initiator codon (ATG), the resultant mutant allele cannot initiate the translation of PTH mRNA into the prepro-PTH protein, and the translocation of the PTH peptide through the ER prior to secretion [32].

4.2 GCM2 gene

Familial isolated hypoparathyroidism-2 (FIH2) is caused by homozygous mutation in the glial cells missing transcription factor-2 gene (GCM2) on chromosome 6p24. Patients with familial isolated hypoparathyroidism-2 (FIH2) usually present with seizures, caused by hypocalcemia, in early life. Serum parathyroid hormone (PTH; 168450) levels are low to undetectable. GCM2 is a transcription factor whose expression is restricted to the parathyroid glands [30]. Biallelic loss-of-function mutations in exons 2, 3, and 5 of GCM2 result in hypoparathyroidism. Mutations in GCM2 exons 2 and 3 (encoding DNA binding and transactivation domain 1) lead to impaired protein synthesis and stability and autosomal recessive transmission of congenital hypoparathyroidism, whereas those in exon 5 (encoding transactivation domain 2) lead to mutations with a dominant negative effect and autosomal dominant transmission of this disorder. Expression of GCM2 occurs immediately after specification of parathyroid cells and is dependent on the normal transcriptional function of the mutated gene, GATA3, in patients with Barakat syndrome. Gain-of-function mutations of GCM2 are associated with hyperparathyroidism [33].

5. Disorders of peripheral unresponsiveness to PTH

5.1 Pseudohypoparathyroidism Type 1a (PHP1a) and Albright Hereditary Osteodystrophy (AHO)

Pseudohypoparathyroidism type 1a (PHP1a), Albright hereditary osteodystrophy (AHO), and Pseudopseudohypoparathyroidism (PPHP) are caused by heterozygous inactivating mutations of the GNAS gene, which encodes the Gsα subunit necessary for peptide hormone signal transduction of GCPRs. Patients with PHP1a have end-organ resistance to PTH with hypocalcemia, hyperphosphatemia, and elevated PTH levels [34]. In addition, patients present with short stature, craniofacial anomalies, shortened fingers, and short fourth and fifth metacarpals characterized by heterotopic ossifications called Albright hereditary osteodystrophy. Patients show resistance to hormones whose receptors bind to Gs such as GHRH, TSH, gonadotropin, calcitonin, and hypothalamic neurotransmitters. PHP1a patients also have neurocognitive deficiency and obesity reflecting the effect of Gαs in the brain. The reason for these two different manifestations of the same gene defect is due to a complex genomic imprinting mechanism that controls the transcription of the GNAS gene [35, 36]. Patients with a GNAS mutation on a maternal allele will develop a more severe form of Gsα deficiency with hormone resistance (PHP1a), whereas patients with identical mutations on the paternal GNAS allele will have a milder condition with normal hormone response (PPHP) [37].

5.2 Parathyroid-hormone-related peptide receptors PTHrP

The type 1 PTH receptor (PTHR1) is activated by PTH and parathyroid-hormone-related peptide (PTHrP) and mediates PTH effects in bone and kidney. Mutations in PTHR1 have been reported in two types of skeletal dysplasia with different clinical manifestations, Metaphyseal chondrodysplasia, Murk Jansen type, a dominant disorder resulting from a gain-of-function mutations and accelerated chondrocyte differentiation, and short-limbed dwarfism in Blomstrand chondrodysplasia, a recessive lethal disorder resulting from loss-of-function mutations [38].

5.3 Calcium sensing receptor (CASR)-related disease

The synthesis and secretion of PTH by parathyroid gland chief cells are regulated by Ca^{2+} concentrations acting through the CaSR that either enhance (when Ca^{2+} concentrations are low) or repress (when they are high) transcription of PTH and secretion of PTH; Ca^{2+} values also modulate the rate of chief cell proliferation, a response also mediated by the CaSR. The extracellular calcium-sensing receptor (CaSR) is a G-protein-coupled receptor (GPCR) that is mostly expressed in the parathyroid and kidneys. CaSR allows regulation of parathyroid hormone (PTH) secretion and renal tubular calcium re-absorption in response to changes in extracellular calcium concentrations [39]. There are calcium-sensing receptor (CASR)-related diseases associated with hypercalcemic and hypocalcemic disorders. While inactivating CaSR mutations are associated with hypercalcemic disorders of familial benign hypercalcemia (FBH), neonatal severe primary hyperparathyroidism (NSHPT), adult primary hyperparathyroidism, and autoimmune hypocalciuric hypercalcemia (AHH); activating CaSR mutations result in autosomal-dominant hypocalcemia with hypercalciuria (ADHH) and Bartter-like syndrome. Furthermore, CaSR auto-antibodies have been found in patients with hypercalcemia or hypocalcemia.

CaSR abnormalities are associated with four hypercalcemic disorders, which are familial benign hypercalcemia (FBH), neonatal severe primary hyperparathyroidism (NSHPT), adult primary hyperparathyroidism, and autoimmune hypocalciuric hypercalcemia (AHH). CaSR abnormalities are associated with three hypocalcemic disorders, which are autosomal-dominant hypocalcemic hypercalciuria (ADHH, Bartter syndrome type V (i.e., ADHH with a Bartter-like syndrome)), and a form of autoimmune hypoparathyroidism (AH) due to CaSR autoantibodies [40].

5.4 Hyperparathyroidism

Pituitary adenomas occur in isolation or may be part of a genetic syndrome, such as multiple endocrine neoplasia type 1 (MEN1) or McCune-Albright syndrome. MEN1 is an inherited tumor syndrome characterized by glandular hyperplasia and benign or malignant neoplasms occurring in two or more endocrine glands, classically the parathyroids, pituitary, and neuroendocrine pancreas [41]. Patients are also at risk for developing adrenocortical tumors, pheochromocytoma (PHEO), extra-abdominal neuroendocrine tumors, benign tumors of the skin (angiofibromas, lipomas, collagenomas, lipomas), and central nervous system tumors (meningiomas and ependymomas). The MEN1 gene is a tumor suppressor gene that encodes a nuclear protein, which plays a role in transcriptional regulation, genome stability, cell division, and proliferation. Most patients with MEN1 inherit the mutation from an affected parent, but about 10% of individuals with MEN1 represent de novo. An inherited heterozygous germline MEN1 mutation is insufficient to induce tumor formation. Therefore, a somatic mutation (the second hit) in the wild-type MEN1 allele is required to cause disease. Primary Hyperparathyroidism (PHPT) is the most common and initial endocrine manifestation of MEN1. MEN1 patients have multi-gland parathyroid hyperplasia rather than single gland adenomas. Most children with MEN1-related PHPT are asymptomatic, and nephrolithiasis is the major hallmark of symptomatic disease [42].

MEN2A is characterized by medullary thyroid carcinoma (MTC), PHEO, and primary hyperparathyroidism. MEN2A is caused by heterozygous gain-of-function mutations in the RET proto-oncogene. More than 90% of RET mutation carriers have a high penetrance for developing thyroid cancer. MEN 2B is characterized by clinically aggressive MTC, pheochromocytoma, a Marfanoid body habitus, mucosal neuromas, and intestinal tumors.

Pathogenic variants involving cysteine codons 609, 618, and 620 in exon 10 of RET are associated with MEN2A, while the RET germline pathogenic variant p.Met918Thr is associated with MEN2B only. More than 90% of RET mutation carriers have a high penetrance for developing thyroid cancer. Most MEN2B cases arise as a result of a de novo mutation, with the child having unaffected parents [43].

6. Conclusion

Although genetic disorders are not the most common cause of hyper/hypoparathyroidism, molecular analyses help us understand mutations in genes involved in calcium and phosphorus metabolism. Furthermore, clarifying the genetic etiopathogenesis of hyper/hypoparathyroidism may contribute to management, prevention of comorbidities, and genetic counseling.

Author details

Ayça Dilruba Aslanger
Istanbul Faculty of Medicine, Medical Genetics Department, Istanbul University, Istanbul, Turkiye

*Address all correspondence to: aaslanger@yahoo.com

IntechOpen

© 2022 The Author(s). Licensee IntechOpen. This chapter is distributed under the terms of the Creative Commons Attribution License (http://creativecommons.org/licenses/by/3.0), which permits unrestricted use, distribution, and reproduction in any medium, provided the original work is properly cited.

References

[1] Blaine J, Chonchol M, Levi M. Renal control of calcium, phosphate, and magnesium homeostasis. Clinical Journal of the American Society of Nephrology. 2015;**10**:1257-1272

[2] Favus MJ, Goltzman D. Regulation of calcium and magnesium. In: Rosen CJ, editor. Primer on the Metabolic Bone Diseases and Disorders of Mineral Metabolism. 7th ed. Washington DC: The American Society of Bone and Mineral Metabolism; 2008. pp. 104-108

[3] Mannstadt M, Bilezikian JP, Thakker RV, Hannan FM, Clarke BL, Rejnmark L, et al. Hypoparathyroidism. Nature Reviews. Disease Primers. 2017;**3**:17055

[4] Bilezikian JP. Hypoparathyroidism. The Journal of Clinical Endocrinology and Metabolism. 2020;**105**:6

[5] Cianferotti L, Marcucci G, Brandi ML. Causes and pathophysiology of hypoparathyroidism. Best Practice & Research. Clinical Endocrinology & Metabolism. 2018;**32**(6):909-925

[6] McDonald-McGinn DM, Sullivan KE, Marino B, Philip N, Swillen A, Vortsman JA, et al. 22q11.2 deletion syndrome. Natural Review Diseases Primers. 2015;**1**:15071

[7] Gong W, Gottlieb S, Collins J, Blescia A, Dietz H, Goldmuntz E, et al. Mutation analysis of TBX1 in non-deleted patients with features of DGS/VCFS or isolated cardiovascular defects. Journal of Medical Genetics. 2001;**38**(12):E45

[8] Bassett AS, McDonald-McGinn DM, Devriendt K, Digilio MC, Goldenberg P, Habel A, et al. International 22q11.2 Deletion Syndrome Consortium. Practical guidelines for managing patients with 22q11.2 deletion syndrome. Journal of Pediatrics. 2011;**159**:332

[9] Al-Jenaidi F, Makitie O, Grunebaum E, Sochett E. Parathyroid gland dysfunction in 22q11.2 deletion syndrome. Hormone Research. 2007;**67**(3):117-122

[10] Cuneo BF, Langman CB, Ilbawi MN, Ramakrishnan V, Cutilletta A, Driscoll DA. Latent hypoparathyroidism in children with conotruncal cardiac defects. Circulation. 1996;**93**(9): 1702-1708

[11] Brauner R, Le Harivel de Gonneville A, Kindermans C, Le Bidois J, Prieur M, Lyonnet S, et al. Parathyroid function and growth in 22q11.2 deletion syndrome. The Journal of Pediatrics. 2003;**142**:504-508

[12] Cheung EN, George SR, Costain GA, Andrade DM, Chow EW, Silversides CK, et al. Prevalence of hypocalcaemia and its associated features in 22q11·2 deletion syndrome. Clinical Endocrinology. 2004;**81**:190-196

[13] Sullivan KE. Chromosome 22q11.2 deletion syndrome and DiGeorge syndrome. Immunological Reviews. 2019;**287**:186-201

[14] Issekutz KA, Graham JM, Prasad C, Smith IM, Blake KD. An epidemiological analysis of CHARGE syndrome: Preliminary results from a Canadian study. American Journal of Medical Genetics. Part A. 2005;**133A**:309-317

[15] Vissers LE et al. Mutations in a new member of the chromodomain gene family cause CHARGE syndrome. Nature Genetics. 2004;**36**:955-957

[16] Zentner GE, Hurd EA, Schnetz MP, Handoko L, Wang C, Wang Z, et al. CHD7 functions in the nucleolus as a positive regulator of ribosomal RNA biogenesis. Human Molecular Genetics. 2010;**19**(18):3491-3501

[17] Barakat AJ, Raygada M, Rennert OM. Barakat syndrome revisited. American Journal of Medical Genetics. Part A. 2018;**176**(6):1341-1348

[18] Van Esch H, Groenen P, Nesbit MA, Schuffenhauer S, Lichtner P, Vanderlinden G, et al. GATA3 haplo-insufficiency causes human HDR syndrome. Nature. 2000;**406**(6794): 419-422

[19] Chenouard A, Isidor B, Allain-Launay E, Moreau A, Le Bideau M, Roussey G. Renal phenotypic variability in HDR syndrome: Glomerular nephropathy as a novel finding. European Journal of Pediatrics. 2013;**172**(1):107-110

[20] Kamezaki M, Kusaba T, Adachi T, Yamashita N, Nakata M, Ota N, et al. Unusual proliferative glomerulonephritis in a patient diagnosed to have hypoparathyroidism, sensorineural deafness, and renal dysplasia (HDR) syndrome with a novel mutation in the GATA3 gene. Internal Medicine. 2017;**56**(11):1393-1397

[21] Padidela R, Kelberman D, Press M, Al-Khawari M, Hindmarsh PC, Dattani MT. Mutation in the TBCE gene is associated with hypoparathyroidism-retardation-dysmorphism syndrome featuring pituitary hormone deficiencies and hypoplasia of the anterior pituitary and the corpus callosum. The Journal of Clinical Endocrinology and Metabolism. 2009;**94**(8):2686-2691

[22] Sanjad SA, Sakati NA, Abu-Osba YK, Kaddoura R, Milner RD. A new syndrome of congenital hypoparathyroidism, severe growth failure, and dysmorphic features. Archives of Disease in Childhood. 1991;**66**(2):193-196

[23] Diaz GA, Gelb BD, Ali F, et al. Sanjad-Sakati and autosomal recessive KennyCaffey syndromes are allelic: Evidence for an ancestral founder mutation and locus refinement. American Journal of Medical Genetics. 1999;**85**(1):48-52

[24] Gordon RJ, Levine MA. Genetic disorders of parathyroid development and function. Endocrinology and Metabolism Clinics of North America. 2018;**47**(4):809-823

[25] Stolarski B, Pronicka E, Korniszewski L, Pollak A, Kostrzewa G, Rowińska E, et al. Molecular background of polyendocrinopathy-candidiasis-ectodermal dystrophy syndrome in a Polish population: Novel AIRE mutations and an estimate of disease prevalence. Clinical Genetics. 2006;**70**(4):348-354

[26] Peterson P, Peltonen L. Autoimmune polyendocrinopathy syndrome type 1 (APS1) and AIRE gene: New views on molecular basis of autoimmunity. Journal of Autoimmunity. 2005;**25**(Suppl):49-55

[27] Root AW. Disorders of mineral metabolism: Normal homeostasis. In: Sperling MA, editor. Pediatric Endocrinology. 5th ed. Philadelphia: Saunders Elsevier; 2020. pp. 234-236

[28] Canaff L, Guarnieri V, Kim Y, Wong BYL, Nolin-Lapalme A, Cole DEC, et al. Novel glial cells missing-2 (GCM2) variants in parathyroid disorders. European Journal of Endocrinology. 2022;**186**(3):351-366

[29] Muroya K, Hasegawa T, Ito Y, Nagai T, Isotani H, Iwata Y, et al. GATA3 abnormalities and the phenotypic

spectrum of HDR syndrome. Journal of Medical Genetics. 2001;**38**:374-380

[30] Guan B, Welch JM, Sapp JC, et al. GCM2-activating mutations in familial isolated hyperparathyroidism. American Journal of Human Genetics. 2016;**99**(5):1034-1044

[31] Cinque L, Sparaneo A, Penta L, et al. Autosomal dominant PTH gene signal sequence mutation in a family with familial isolated hypoparathyroidism. The Journal of Clinical Endocrinology and Metabolism. 2017;**102**(11):3961-3969

[32] Lee JH, Davaatseren M, Lee S. Rare PTH gene mutations causing parathyroid disorders: A review. Endocrinology Metabolism (Seoul). 2020;**35**(1):64-70

[33] Riccardi A, Aspir T, Shen L, Kuo CL, Brown TC, Korah R, et al. Analysis of activating GCM2 sequence variants in sporadic parathyroid adenomas. The Journal of Clinical Endocrinology and Metabolism. 2019;**104**(6):1948-1952

[34] Weinstein LS, Liu J, Sakamoto A, Xie T, Chen M. Minireview: GNAS: Normal and abnormal functions. Endocrinology. 2004;**145**(12):5459-5464

[35] Turan S, Bastepe M. GNAS spectrum of disorders. Current Osteoporosis Reports. 2015;**13**(3):146-158

[36] Turan S, Bastepe M. The GNAS complex locus and human diseases associated with loss-of-function mutations or epimutations within this imprinted gene. Hormone Research in Pædiatrics. 2013;**80**(4):229-241

[37] Lebrun M, Richard N, Abeguile G, David A, Coeslier DA, Journel H, et al. Progressive osseous heteroplasia: A model for the imprinting effects of GNAS inactivating mutations in humans. The Journal of Clinical Endocrinology and Metabolism. 2010;**95**(6):3028-3038

[38] Mantovani G, Elli FM. Inactivating PTH/PTHrP signaling disorders. Frontiers of Hormone Research. 2019;**51**:147-159

[39] Riccardi D, Brown EM. Physiology and pathophysiology of the calcium-sensing receptor in the kidney. American Journal of Physiology. Renal Physiology. 2010;**298**(3):F485-F499

[40] Thakker RV. Calcium-sensing receptor: Role in health and disease. Indian Journal of Endocrinology Metabolism. 2012;**16**(Suppl 2):S213-S216

[41] Thakker RV, Newey PJ, Walls GV, et al. Clinical practice guidelines for multiple endocrine neoplasia type 1 (MEN1). Journal of Clinical Endocrinology and Metabolism. 2012;**97**(9):2990-3011

[42] Kamilaris CDC, Stratakis CA. Multiple endocrine neoplasia type 1 (MEN1): An update and the significance of early genetic and clinical diagnosis. Frontiers in Endocrinology (Lausanne). 2019;**11**(10):339

[43] Waguespack SG, Rich TA, Perrier ND, Jimenez C, Cote GJ. Management of medullary thyroid carcinoma and MEN2 syndromes in childhood. Nature Reviews. Endocrinology. 2011;**7**(10):596-607

Chapter 6

Evaluation of Parathyroid Pathophysiology via Cell Distribution and Expression Patterns

Beyza Goncu

Abstract

The parathyroid tissue is composed of the chief, oxyphil, and water-clear cells. The cell type in each parathyroid gland is highly heterogeneous between different pathologies. The parathyroid oxyphil cells are markedly increased in secondary hyperparathyroidism due to chronic kidney diseases. These cells include more eosinophil than oxyphil cells, but they are closer in size to the chief cells. Studies reported that the oxyphil cells are derived from chief cells, and this presents another cell type that occurs as "transitional oxyphilic cells." As is known, calcium-sensing receptor (CaSR) is expressed abundantly in the chief cells. Expression of CaSR is elevated in disparate parathyroid tissues, which is possibly related to differential expression levels of parathyroid-specific transcription factors including *GCM2* (Glial Cells Missing Transcription Factor 2), *MAFB* (V-maf musculoaponeurotic fibrosarcoma oncogene homolog B), *GATA3* (GATA Binding Protein 3), *RXR* (The retinoid X receptor), and even *VDR* (Vitamin D Receptor). The pathways that connect CaSR to parathyroid cell proliferation are precisely not known yet. Evaluation of oxyphil and chief cells of parathyroid glands and their differential expression patterns are important to understand the parathyroid function and its behavioral changes due to related diseases. This chapter presents a summary of the current literature on the cell type distribution of parathyroid and pathophysiology by comparing the expression patterns.

Keywords: parathyroid gland, hyperplasia, adenoma, oxyphil cells, chief cells, water-clear cells

1. Introduction

Restoration of calcium levels is essential for the human body, which leads to proper function such as enzyme activity, hormone secretion, neurological stimulation, and/or muscle function. Calcium homeostasis is in relationship with different tissues and structures such as bone, parathyroid gland, and kidney. Regulation of calcium-homeostasis-related events begins with rapid changes in the parathormone (PTH) release from parathyroid gland cells. Hypocalcemic or hypercalcemic

IntechOpen

responses are controlled by a specific receptor that senses the changes in serum calcium levels [1]. If calcium levels decrease, the mechanism provides a rapid increase in the PTH level to maintain proper calcium levels. Once calcium is balanced to the normal values, continual calcium release suppresses the PTH through a negative-feedback mechanism. Major and pulsatile events occur during the process itself through calcium-sensing receptors (CaSRs) [2]. This G-protein-coupled receptor is a unique part of the parathyroid gland function to monitor changes in serum calcium levels. This process is carried out by two main mechanisms which are, respectively, stimulation of the kidneys and intestine to increase the absorption of calcium and stimulation of the bones to release calcium into the blood [3, 4].

PTH mRNA expression levels are suppressed by the 1-25OH$_2$D modulation, despite CaSR expression with its effect on the modulation of the PTH release. Ritter et al. report that PTH release decreased by 1–25(OH)$_2$D, 1 hydroxy-vitamin D, and 25(OH) D in mice parathyroid cells in culture [5–11]. Earlier in this study, Kim et al. showed a lack of upregulation of PTH transcription in the Vitamin D Receptor (VDR) knockout mice, and lack of suppression of PTH transcription by 1,25(OH)$_2$D administration [12]. Consistent with both researcher and current literature, the 1–25(OH)$_2$D reduces the PTH mRNA expression and has an antiproliferative effect on parathyroid cells from uremic rats, subsequently enhancing the VDR expressions. Acute changes in the calcium levels of sHPT patients enhance the CaSR and Klotho expressions by Vitamin D [5, 13].

The location of the parathyroid glands is on the posterior side of the thyroid gland. Mostly there are four glands and rarely supernumerary glands [14]. Starting from the early phases of embryogenesis, the pharyngeal pouches give rise to parathyroid glands along with many organs [15, 16]. The third and the fourth pharyngeal pouches particularly emerge as parathyroid glands [17]. Inferior and superior parathyroid glands develop with thymus and a portion of the thyroid respectively [15]. Different cell adhesion after separation from the pharyngeal pouch modulates the formation of parathyroids, while the thymus continues to migrate above the heart [16]. Separation of the superior parathyroid glands is carried out during the seventh week of the development and after detaching the pharyngeal wall, the parathyroid fuses with the posterior surface of the thyroid [14].

The parathyroid glands are derived from the endoderm during embryogenesis. Endoderm consists of a high amount of actin fibers that provide formation and expansion to the ectoderm. Numerous signaling molecules and proteins maintain the progression of morphogenesis. Particularly, Glial cell missing 2 (*GCM2*), Homeobox A3 (*HOXA3*), Forkhead box protein N1 (*FOXN1*), Eyes Absent Homolog 1 (Transcriptional Coactivator And Phosphatase 1—*EYA1*), T-Box Transcription Factor 1 (*TBX1*), GATA binding protein 3 (*GATA3*), Paired box 1 (*PAX1*), and Paired box 9 (*PAX9*) genes are the leading transcription factors that initiate the parathyroid morphogenesis [16, 17]. Among them, *GCM2* is particularly important as the early transcription factor of parathyroid formation [18]. The developmental stage includes another important transcription factor: V-maf musculoaponeurotic fibrosarcoma oncogene homolog B (*MAFB*) [19].

In order to understand the development of the parathyroid gland, extensive research on murine models is performed. To identify the involving molecules and signaling processes, researchers have proven most of the downstream and upstream mediators. The remaining unknown factors remain to be elucidated. In current practice, the known factors involving the development and maturation of the parathyroid glands are evaluated in murine models and related parathyroid disorders. This chapter summarizes the molecular findings based on the cell type distribution of parathyroid and pathophysiology up to the present.

2. Cellular distribution of parathyroid gland

The vascular structure of the parathyroid glands is separated from the outside by a fibrous thin capsule. Such capsules contain thin fibrous bands that disperse toward the inner parts by fibrous structures. This provides access to the blood vessels, lymphatics, and nerves to provide nutrients to the tissue [20, 21]. Macroscopic appearances may vary depending on cell type, cell/organelle amount, blood supply status, and the amount of fat content of the tissue [1, 22].

In literature, cell-type distribution in the parathyroid tissue has been reported based on either two or three different cell types. In some reports, it has been stated as two: oxyphil and chief cells [14], while another group of researchers has elucidated histological evidence that there are water-clear cells besides oxyphil and chief cells [23]. Besides, there is another group of cell types that are observed with disease and/or age-related changes, which are not yet classified as a specific cell type. Although this type is not yet classified, the transitional chief cells are present and particularly reported as *chief cell-to-oxyphil cell transdifferentiated cell* group [23, 24]. These four types of cells do not have distinctive markers yet. Their variations in size and number per tissue originate from different parathyroid pathophysiology, which is the main unknown aspect of parathyroid cell biology. Hence, this chapter prioritizes the three main cell types in the parathyroid gland.

2.1 Water-clear cells

Water-clear cells contain many glycogen granules in their cytoplasm and are rarely seen. These cells are clinically encountered in the tissues of patients with secondary hyperparathyroidism (sHPT) and primary hyperparathyroidism (pHPT) [25]. In 1992, Emura's study on rabbit parathyroid tissue stated that water-clear cells contained a large number of vacuoles in the parathyroid tissue sections, which were observed with electron microscopy and were found scattered around the chief cells. Water-clear cells reside in between the perivascular space and the basal lamina that have been observed to have a desmosomal connection with chief cells [26]. In 2013, Ezzat et al. showed that these cells were not associated with PTH secretion and serum calcium levels. In addition, they reported that only 0.3% of pHPT cases had "water-clear cell hyperplasia" or "water-clear cell adenomatosis" [27]. A recent report presented that the water-clear cell hyperplasia ratio is less than 1% of all pHPT cases [28]. Distinct features or changes in the water-clear cell accumulation in parathyroid tissue or relation with disorders remain to be elucidated.

2.2 Chief cells

The chief cells constitute the cellularly dominant cell type of parathyroid tissue, and it was described by Baker in 1942 [29]. Baker describes these cells as the dark "primary cells" with distinctive cytoplasmic structures. Their rod-shaped mitochondria are homogeneously distributed throughout the cytoplasm [20]. In addition, Trier reported that he did not observe the dark and pale stained oxyphil cells, neither with light nor with electron microscopy, which Baker mentioned in his notes in 1942.

The chief cells are usually rich in intracellular fat content. Cells are supported by a thin connective tissue and accordingly are located more closely to the capillaries [30]. Although the chief cells may contain more than one nuclei (multinucleic cells), the nuclear matrix structures are densely arranged [31]. Chief cells are considered the main cell type, and cells are mostly spherical or oval-shaped with long nuclei and narrow cytoplasm [32]. The knowledge about the cell shapes of the chief cells is mostly obtained from histological examinations. Besides, observations by using live imaging or confocal microscopy are either absent or limited. Since it has been reported that single chief cell diameters are 0.2 μm wide according to electron microscope images [33, 34] and 6–8 μm histologically [23]. The agranular membrane structure of the Golgi body of the chief cells was first visualized in 1957 [35]. Due to the eosinophilic cytoplasm, they may appear dark or light in color at the time of staining.

To date, the parameters such as age, disease history, and drug use, which are the definitive features of parathyroid function, may affect cytoplasm amount or changes in the cytoplasm content or nucleus size for all processes [32, 36, 37]. In brief, functional activity and cytoplasm content are the two related main chief cell behaviors of the parathyroid. The most distinctive feature of the chief cells is that they contain a large number of secretory vesicles. These membrane-covered vesicles mostly contain PTH [38].

As is known, the main function depends on the chief cells, which are responsible cell type for PTH release. These cells play an important role in calcium homeostasis, thanks to the CaSR expression on its surface. The receptor detects the low amount of extracellular calcium changes and releases the appropriate amount of PTH to balance the calcium in the blood [23, 39].

2.3 Oxyphil cells

Oxyphil cells are the cells with well-circumscribed eosinophilic cytoplasm and pycnotic nuclei [1, 20, 39]. Between 1952 and 1953, Parade compared monkey, equine, and human parathyroid tissues and reported that the size and number of mitochondria in oxyphil cells varied between species. In addition, the variability in the number of mitochondria in human oxyphilic cells is also associated with age [23, 33, 34]. As of note, rat parathyroid tissue does not contain oxyphil, oxyphil-like, or mitochondria-rich cell groups [32]. In 1958, Trier observed that some of the oxyphilic cells were stained “dark” and some were “pale.” He described pale-stained oxyphilic cells as having “low” mitochondrial content, and dark-staining oxyphil cells as having “excessive” mitochondrial content [20]. Both studies have confirmed the outcome of oxyphil cells.

In 1981, Allen and Thorburn examined the activity of oxyphil cells in abnormal parathyroid tissue of 114 patients with sHPT due to chronic kidney disease. The absence and presence of oxyphil cells in human parathyroid tissues were evaluated in this retrospective study, in which histological evaluation was associated with clinical practice. They reported that hyperparathyroidism was seen in more than one parathyroid tissue in 55% of the cases, and adenoma was found in one of the four parathyroid glands in 69% of the cases. They also reported that oxyphil cells were found in 91% of the cases, and the number of oxyphilic cells was positively correlated with serum calcium level [40].

In 1990, Suzuki et al. reported oxyphil cell function in 148 parathyroid tissues from patients who are taking hemodialysis. They calculated for each tissue by proportioning the area occupied by oxyphilic cells by morphometric measurements concerning the total parathyroid cross-sectional area (oxyphil cell area/total area). Additionally, it has been reported that serum PTH levels do not have a statistically significant relationship with age and the dialysis duration; however, the total tissue size is positively correlated with PTH release. Based on the results, they concluded that the oxyphil cell/total tissue area was not effective in PTH release in patients with chronic renal failure [41].

In 1996, Tanaka et al. used 22 sHPT tissues to understand oxyphil cell function. The study reported that 10 of these tissues had oxyphil cells and the mRNA expression of PTH was found to increase. Then, performed heterotransplantation in mice to determine oxyphilic cell function by evaluating serum PTH levels. As a result, the change in PTH levels was positively correlated with transplanted tissue size, not the cell number or type [42].

Despite the reported studies, there is no definite information about the exact function of oxyphil cells, but this question was clarified in many ways by Ritter and Brown et al. [5, 6, 23, 39]. Histologically, increased eosinophilic content from the chief cell through oxyphil cell, suggesting that oxyphil cells are formed by "transition." As evidence, the oxyphil cells have been shown to express PTH [42] and *GCM2*, which is a parathyroid-specific transcription factor [18] and has a role in parathyroid tissue development. It has been observed that oxyphil cells are more numerous in patients with chronic renal failure, and the amount of oxyphil cells is much higher than in the tissues of healthy individuals [43, 44]. Although studies have shown that oxyphil cells express the parathormone-dependent protein (PTHrP) [45, 46] and that protein is involved in PTH release [42], the amount/mechanism of PTH release is not yet known. The proposed function of the PTHrP on parathyroid cells may be responsible for autocrine or paracrine signaling while affecting PTH release or parathyroid maturation [23].

On the one hand, the CaSR expression levels of oxyphil cells are statistically found higher than other parathyroid cells. On the second hand, there was no significant difference in Vitamin D Receptor (VDR) expression [5, 39]. The higher mitochondrial content of oxyphil cells indicates that energy requirements are much higher than in chief cells. Mitochondria are also responsible for VDR function. One study by Ritter et al. elucidated that 25-hydroxyvitamin D-1α-hydroxylase (1αOHase) is highly expressed and this is the inactive form of Vitamin D [5]. In addition, the amount of 1αOHase in human parathyroid tissue was directly proportional to calcium levels. The study reported that calcimimetic therapy in patients with chronic renal failure causes a significant increase in the amount of 1αOHase in oxyphil cells [5, 6].

A recent commentary also highlighted Ritter's findings after 5 years. Metabolic changes of parathyroid tissue are significantly affected by changes in tissue volume and/or cell type, cell count rate due to drug use, or changes in serum calcium level. Concomitant CaSR induction and its persistence are also known to affect the expression profile [47]. Thus, in terms of PTH expression, it has been clarified that oxyphil cells contain more PTH than chief cells. Some of these data also confirm the findings of Allen and Thorburn in 1981 [40].

3. Cellular variations in parathyroid research and related diseases

Calcium or di-/trivalent cations induce the activation of CaSR, which triggers the PTH release [48]. Definitive research studies have demonstrated the outcomes from direct or indirect approaches so far. The comparisons and the evaluations were mainly performed with the diseased tissues, not the healthy parathyroids due to the difficulties in retrieval processes. There is only a limited number of papers that compare/evaluate healthy parathyroid tissue expression profiles of cellular content. Particularly, the location and the size of the parathyroid gland make it difficult to obtain from healthy individuals. Researchers, surgeons, and physicians reported different approaches to finding and/or distinguishing the parathyroid tissue during thyroid operation [49–51]. This challenge still exists, and suggested techniques such as near-infrared autofluorescence [52] are not readily available for the use of numerous surgeons. Essentially, it is a fact that even if healthy tissue is donated, it will take a long time to reach the appropriate sample size required to conduct various studies. Therefore, a limited number of healthy parathyroid tissues either used or to be used in studies for comparisons.

The further part of this chapter continues with parathyroid tissue from the primary and secondary hyperparathyroidism patients (adenoma and hyperplasia tissues respectively) were compared according to their cellular content. The changes in their expression profiles were evaluated with different stages of the related diseases.

Starting 25 years ago, most of the papers evaluated the CaSR, PTH, proliferation markers, and transcription factors expression changes by mRNA and/or protein levels (mostly immunohistochemistry, western blot, ELISA methods). The clinical characteristics and the cell content of the parathyroid tissue were compared by Yamaguchi et al. in 1997. Samples are retrieved from patients who have secondary hyperparathyroidism and cell proliferation specific marker PCNA (proliferating cell nuclear antigen) expression was compared between normal, adenoma, and hyperplasia parathyroid tissues. This study divided the cell content in each tissue group including dark-stained chief cells, clear chief cells, vacuolated chief cells, transitional oxyphil cells, and oxyphil cells [53]. This divided cell type classification was very similar to the report by Trier in 1958 [20]. Among 27 out of 40 normal parathyroid tissues were found positive for PCNA, and no correlation between age and expression levels is observed. Cell content was reported as clear-chief>dark-chief>oxyphil cells, respectively. However, the normal parathyroid tissue was obtained from thyroid cancer patients, and a definite interpretation should not be made without ignoring this situation. Adenoma tissue showed remarkably higher PCNA expression levels, and the cell content included mostly clear chief cells and less commonly transitional oxyphil cells. In 129 parathyroid hyperplasia tissue samples, all of the divided cell types of this study were found distinguishable. The PCNA expression was found significantly higher in the nodular type from the glandular structure of the parathyroid. The authors concluded that clear chief cells are the most proliferative cell group in all samples, and the dark chief cell group was found the lowest proliferative group [53]. Yamaguchi's study alone is one of the rare studies that evaluate the highest number of parathyroid tissues in its related field.

In 2000, Corbetta et al. demonstrated that 27 parathyroid adenoma tissues contain only chief cells. Cell isolation, cultivation under different calcium concentrations, PTH levels, and CaSR expression levels were evaluated. Additionally, it is stated that there was no correlation between different calcium sensitivity and pathology. Furthermore, PTH and CaSR mRNA and protein levels were significantly reduced when compared with normal tissue, and the authors concluded that defective calcium sensing occurs in abnormal parathyroid tissue [54]. This study may be considered as a starting point for understanding the defects in the sensing mechanism of the CaSR. Failure to obtain the expected changes in the PTH level under different calcium concentrations should not be interpreted as a resistance mechanism.

DOI: http://dx.doi.org/10.5772/intechopen.106228

In 2006, Brown et al. stated that three different Vitamin D prodrugs, which are lacking side-chain hydroxyl groups, were treated with bovine parathyroid cells and showed that PTH levels decreased based on the related concentrations. The prodrugs have different affinities to the VDR; however, utilization of the VDR decreases PTH synthesis for treatment of secondary hyperparathyroidism [7]. The inhibition of PTH synthesis was performed in vitro using prodrugs at that time of the work, and this indicates a new aspect of parathyroid research. Continued with Ritter et al. in 2006, by the same research group, competitive VDR binding of the vitamin D analogs was examined. In this report, two main conclusions were included. One is the 25(OH)D_3 has the highest affinity to the VDR among other analogs, and direct action mechanisms through VDR suppress PTH [8]. The PTH regulation versus VDR activation may be explained by a compensatory mechanism model. Although the authors did not exclude other regulatory systems such as the negative feedback mechanism of the PTH [10]. Studies conducted between 2005 and 2011 mostly did not focus on cell-type-specific changes. Instead, the relationships between VDR and PTH were investigated in terms of regulation mechanism, especially in bovine, rat, and other knockout-animal models.

In 2012, Ritter et al. defined the differential mRNA expression of parathyroid glands by cell types. In this paper, histological examination was provided and the size of the chief, oxyphil, and transdifferentiated oxyphil cells was reported. Consistent with previous reports, the high oxyphil cell amount was reported in chronic kidney diseased patients, accordingly higher PTH and CaSR expression was elucidated as well. In addition, oxyphil cells showed *GCM2* expression, which is a specific parathyroid transcription factor. Therefore, in order to understand parathyroid pathophysiology in patients with secondary hyperparathyroidism, oxyphil cell secretomes may help to define their role other than chief cells [23]. This study still has important outcomes that shaped the perspective to a particular point for chief and oxyphil cell features. Future studies including isolation of oxyphil and chief cells separately with the assessment of secretory features will justify each cell type's function.

In 2014, Shi et al. demonstrated a flow cytometric cell sorting of 20 parathyroid adenoma tissues from primary hyperparathyroidism patients. They divided the cells into three distinct populations including the chief, oxyphil, and lymphocytes. The cutting-edge research from Shi et al. provided electron microscopy images of each population and also demonstrated the immunofluorescence staining of CaSR and mRNA expressions of *CASR* and *PTH* by comparing oxyphil and chief cells. At end of this unique study, they reported that oxyphil cells respond to calcium faster than chief cells by releasing higher PTH and did not find differences in their CaSR expression profiles. They also reported that the feature of oxyphil cells provides an important function to the parathyroid tissue as a piece of solid evidence [55].

In 2015, Howson et al. investigated the oxyphil-cell-rich adenoma tissues from primary hyperparathyroidism patients. During the 10-year follow-up period, among obtained 2739 tissues, 91 of the parathyroid adenoma were found oxyphil cell adenoma type. About 80% of these patients were symptomatic and most commonly had higher serum calcium and PTH levels than the classical type of adenoma. On the contrary, the frequency of oxyphilic adenomas was not rare, and patients trend toward a higher rate of morbidity and potential mortality if left untreated [56].

In 2017, two different groups presented water-clear cell-type adenoma and hyperplasia cases. The clinical symptoms of the adenoma patient were unintentionally led and treated for hyperparathyroidism due to the clinical features. However, after surgical removal of the two adenoma tissues from the same patient, the histopathological evaluation showed water-clear cell double parathyroid adenomas [57]. This is followed by another case report that presents primary hyperparathyroidism clinically. Contrary to the clinic, the histopathological results showed enormous unilateral water-clear cell hyperplasia in parathyroid [28]. Both of the cases concluded that despite ultrasonography, biochemical, and clinical follow-up, these extremely rare cases unintentionally misled the physicians [28, 57]. In the same year, another study by Ritter et al. was reported. In that study, parathyroid hyperplasia tissues were retrieved from chronic kidney patients and grouped according to their calcimimetic treatment (cinacalcet versus paricalcitol versus cinacalcet+paricalcitol). The effects of treatment processes on the cell type of the parathyroid tissue were compared with four healthy parathyroid tissues. According to the histopathological evaluation, parathyroid oxyphil cell content was found to increase significantly for the cinacalcet-treated patients [58]. The role of the CaSR activation possibly led to a new comprehension to understand the outcome of conventional treatment with vitamin D analogs or calcimimetics on the cellular composition of the parathyroid.

In 2020, Ding et al. provided a comparison of clinical characteristics and oxyphil cell proportion through 78 patients. In total, 295 parathyroid tissue samples were retrieved from 78 patients who did not have cinacalcet treatment. Clinical characteristics included serum calcium, phosphorus, alkaline phosphatase, age, dialysis duration, and preoperative PTH levels. The samples were divided based on the mean oxyphil cell ratio (high oxyphil cell content and low oxyphil cell content, respectively). Subsequently, etiopathogenesis and histological examinations were evaluated. They reported that preoperative PTH levels of the patients were found lower than the oxyphil cell-rich group [37]. This finding contradicts the previous study by Howson [56]. Furthermore, Ding et al. reported that if parathyroid tissue contains more oxyphil cells, it became lighter than the tissue with fewer oxyphil cells. As of note, weight comparison was performed between parathyroid hyperplasia tissues [37].

At the beginning of 2021, Altinay et al. demonstrated the cellular composition of the hyperplasia and adenoma tissues while comparing normal parathyroid glands. Furthermore, PTH, GATA3, and PAX8 levels were evaluated histologically. As a result, expression of GATA3 and PTH was found more prominent in pathologic parathyroid tissues when compared with normal. Particularly the GATA3 staining has shown positive only for parathyroid, not thyroid tissue. Chief cell amount was high in adenoma and hyperplasia tissues; however, PTH staining was found low when compared with normal tissue. In addition, adenoma displayed less PTH and GATA3 expression histologically [59].

Another study from Rodriguez et al. reported the clinicopathological outcome of the oxyphil cell clusters, which were detected in parathyroid adenoma tissue. The main idea is to define the particular effect of the oxyphil cell content-rich/poor tissue composition and localization. Despite clinic versus histopathological comparisons so far, this study has a similar manner with distinct oxyphil cell subgroups [60]. Rodriguez's team investigated histopathological function with symptomatology, and this could depend on the changes in the percentage of the oxyphil cells. Clinically, observations included age, sex, body mass index, and symptomatic reasons for surgical initiation such as nephrocalcinosis, osteoarticular and/or neuropsychiatric and/or cardiovascular symptoms. Biochemical parameters were as follows: ionic calcium, corrected serum calcium, albumin, PTH, 25-OH Vitamin D, alkaline phosphatase, creatinine, glomerular filtration rate (GFR), and urinary calcium excretion. Additionally, oxhyphil cell percentages were divided into the following four categories: 0–24, 25–49, 50–74, and 75–100% [60]. In terms of variables, this is the most comprehensive evaluation to date associated with clinical (including both biochemical and symptomatic) and parathyroid cell groups. In total 261 parathyroid tissues obtained from 238 patients were used. Eventually, 77% of tissues have less than 25% in the percentage of oxyphil cells and 8% of tissues are greater than 75% in terms of percentage of oxyphil cells. No significant difference was found in terms of biochemical parameters such as calcium, phosphorus, alkaline phosphatase, PTH, and 25-OH vitamin D. In addition, the localization of the adenoma tissues did not show significant changes when compared between inferior and superior glands. Different distributions of the oxyphil cell clusters among varied cell percentages showed no significant changes. Although, increased thyroid nodularity and higher prevalence in cardiovascular symptoms are shown within the less oxyphil cell groups (<25%). Imaging tests also justified a correlation between better localization with increased oxyphil cell group (>75%). Nevertheless, preoperative GFR and urinary calcium excretion significantly worsen the patients' symptoms that were altered in parathyroid tissues containing high oxyphil cells. Findings in the non-pathological tissue samples from normocalcemic patients showed an absence/low level in the oxyphil cells [60]. These conditions are still controversial.

3.1 Exception: parathyroid carcinoma

Parathyroid carcinomas influence the laboratory values, which are similar to those with hyperparathyroidism (primary or secondary). For instance, increased serum PTH levels are around thousands, and a palpable mass may be detected in the neck region. Currently, carcinoma can be observed in any (upper or lower) of the glands and is not considered to have any priority within the four existing glands while the invasion of the adjacent thyroid gland is observed [61]. The histological resemblance of parathyroid adenoma causes a challenge for diagnosis. Observations in parathyroid carcinoma tissue include increased mitotic potential, necrosis, formation of encapsulated structures, and spread from the capsule through adjacent tissues [62].

Studies on parathyroid cancer are gaining attention. A recent study evaluated whether tumor volume and tumor size were associated with disease severity [63]. Another study determined circulating miRNA expression levels as a potential diagnostic biomarker in parathyroid carcinomas [64, 65].

Despite all these findings, in terms of cell type, water-clear, oxyphil, and chief cells do not provide a differential diagnosis. The prime reasons are a rarity, and it is not possible to impose a limitation in terms of cell composition.

4. Conclusion

To further understand the parathyroid gland function and development, studies were carried out by numerous researchers. Indicated contributions already shaped the required future studies for this composite tissue. The parathyroid gland is a relatively small tissue while the behavior and changes in its composition provide a fine balance in terms of its function. Thus, this has enabled us to come a little closer to elucidating the actual mechanisms. Foremost, the mechanisms that maintain certainty in all research results can be listed as the following:

- Parathyroid cells mostly contain three types of cells; chief, oxyphil, and water-clear cells.
- Depending on the circumstances such aging and diseases may affect and increase the oxyphil cell amount.
- The cells that switch between chief through oxyphil cells can affect different mechanisms, which are still uncategorized.
- Categorization of the differentiating cells may explain the influence of autocrine/paracrine effect on tissue behavior.
- So-called "transdifferentiated cells" can be classified as another cell type that will provide a separate diagnosis criterion such as prognosis, degree, or etiology of parathyroid-related-diseases.

Expression patterns of the same transcription factors such as GCM2, MAFB, and RXR are the most difficult part of distinguishing the cell-type-specific features. Even with all the findings in the literature, one question still remains; where does the border of differential cellular diagnosis end? The answer to this question is still unknown.

The role of autocrine and paracrine effects in parathyroid cell differentiation cannot be ignored. Each person's metabolism balances this process individually; therefore, studies should include larger cohorts. More collaborative studies are required between researchers. The lack of oxyphil cells in some murine models is another challenge. This indicates that understanding the cellular composition/regulation of parathyroid tissue behavior is mandatory, particularly for primary human parathyroid tissue cell studies.

The emergence of oxyphilic cells is perhaps a defense mechanism that is developed from the parathyroid tissue against external signalings, which was first reported by Christi in 1955 [31]. However, this is yet to be confirmed. Another remark can be made from the state of immunogenicity, which is considered as a defense mechanism or after pathological disturbances leads to adverse circumstances, and as a result, the oxyphil cells differentiate from chief cells. Nevertheless, several studies have focused on the presence of other immunological markers for parathyroid tissue. Verified studies to date have limited definitions for the expression of immunological markers such as human leukocyte antigens (HLAs) [66–69]. Revealing the possible relationship of immunological and/or defense mechanisms in parathyroid cell composition requires more studies such as different co-culture models. This can be a starting point for future studies.

A few important subjects that have not been finalized yet in terms of parathyroid tissue are:

- Histological studies are contributing to the field; however, more biomarker research is required for the specific differential prognosis of related diseases.
- The inverse relationship between the increased oxyphilic cell and the lighter tissue structure is still not understood, whereas the studies that determine weight through cell composition of the parathyroid tissue could provide paramount importance.

In conclusion, the approaches that have been proposed in the literature paved the way for future research objectives. Differentiating points obtained by comparing the outcomes and the data of different researchers raise new questions. The exact mechanism for basic parathyroid biology requires new *in vitro* and *in vivo* approaches, and mostly, primary tissue culture systems are essential to understand such a mechanism.

Acknowledgements

The author would like to thank Prof. Emrah Yucesan for his valuable feedbacks. This chapter is dedicated to all patients undergoing treatment for parathyroid-related diseases who agreed to participate in the studies from many researchers highlighted in this section. This chapter was supported by the Research Fund Unit of the Bezmialem Vakif University.

Conflict of interest

The authors declare no conflict of interest.

Author details

Beyza Goncu[1,2]

1 Parathyroid Transplantation Laboratory, Organ Transplantation Center, Bezmialem Vakif University Hospital, Istanbul, Turkey

2 Department of Medical Services and Techniques, Vocational School of Health Services, Bezmialem Vakıf University, Istanbul, Turkey

*Address all correspondence to: bgoncu@bezmialem.edu.tr; bsgoncu@gmail.com

IntechOpen

© 2022 The Author(s). Licensee IntechOpen. This chapter is distributed under the terms of the Creative Commons Attribution License (http://creativecommons.org/licenses/by/3.0), which permits unrestricted use, distribution, and reproduction in any medium, provided the original work is properly cited. (cc) BY

References

[1] Baloch ZW, LiVolsi VA. Parathyroid glands, pathology. In: Reference Module in Biomedical Sciences. 3rd ed. Philadelphia, PA, USA: Elsevier; 2014. p. 474

[2] Chiavistelli S, Giustina A, Mazziotti G. Parathyroid hormone pulsatility: Physiological and clinical aspects. Bone Research. 2015;**3**:14049

[3] Hall JE. Guyton and Hall Textbook of Medical Physiology. 13th ed. Vol. xix. Philadelphia, PA: Elsevier; 2016. p. 1145

[4] Lofrese JJ, Lappin SL. Physiology, Endocrine, Parathyroid. Treasure Island (FL): StatPearls; 2018

[5] Ritter CS, Haughey BH, Armbrecht HJ, Brown AJ. Distribution and regulation of the 25-hydroxyvitamin D3 1alpha-hydroxylase in human parathyroid glands. The Journal of Steroid Biochemistry and Molecular Biology. 2012;**130**(1-2):73-80

[6] Brown AJ, Ritter CS, Weiskopf AS, Vouros P, Sasso GJ, Uskokovic MR, et al. Isolation and identification of 1alpha-hydroxy-3-epi-vitamin D3, a potent suppressor of parathyroid hormone secretion. Journal of Cellular Biochemistry. 2005;**96**(3):569-578

[7] Brown AJ, Ritter CS, Knutson JC, Strugnell SA. The vitamin D prodrugs 1alpha(OH)D2, 1alpha(OH)D3 and BCI-210 suppress PTH secretion by bovine parathyroid cells. Nephrology, Dialysis, Transplantation. 2006;**21**(3):644-650

[8] Ritter CS, Armbrecht HJ, Slatopolsky E, Brown AJ. 25-hydroxyvitamin D(3) suppresses PTH synthesis and secretion by bovine parathyroid cells. Kidney International. 2006;**70**(4):654-659

[9] Brown AJ, Ritter CS. The vitamin D analog 1alpha,25-dihydroxy-2beta-(3-Hydroxypropyloxy) vitamin D(3) (eldecalcitol) is a potent regulator of calcium and phosphate metabolism. Calcified Tissue International. 2011;**89**(5):372-378

[10] Ritter CS, Brown AJ. Direct suppression of Pth gene expression by the vitamin D prohormones doxercalciferol and calcidiol requires the vitamin D receptor. Journal of Molecular Endocrinology. 2011;**46**(2):63-66

[11] Ritter CS, Brown AJ. Suppression of PTH by the vitamin D analog eldecalcitol is modulated by its high affinity for the serum vitamin D-binding protein and resistance to metabolism. Journal of Cellular Biochemistry. 2011;**112**(5):1348-1352

[12] Kim MS, Fujiki R, Murayama A, Kitagawa H, Yamaoka K, Yamamoto Y, et al. 1Alpha,25(OH)2D3-induced transrepression by vitamin D receptor through E-box-type elements in the human parathyroid hormone gene promoter. Molecular Endocrinology. 2007;**21**(2):334-342

[13] Bienaime F, Prie D, Friedlander G, Souberbielle JC. Vitamin D metabolism and activity in the parathyroid gland. Molecular and Cellular Endocrinology. 2011;**347**(1-2):30-41

[14] Rosen RD, Bordoni B. Embryology, Parathyroid. Treasure Island (FL): StatPearls; 2021

[15] Peissig K, Condie BG, Manley NR. Embryology of the parathyroid glands. Endocrinology and Metabolism Clinics of North America. 2018;**47**(4):733-742

[16] Miles B, Srinivasan VN. Embryology, Pharyngeal Pouch. Treasure Island (FL): StatPearls; 2021

[17] Parvari R, Diaz GA, Hershkovitz E. Parathyroid development and the role of tubulin chaperone E. Hormone Research. 2007;**67**(1):12-21

[18] Nonaka D. Study of parathyroid transcription factor Gcm2 expression in parathyroid lesions. The American Journal of Surgical Pathology. 2011;**35**(1):145-151

[19] Naveh-Many T, Silver J. Transcription factors that determine parathyroid development power PTH expression. Kidney International. 2018;**93**(1):7-9

[20] Trier JS. The fine structure of the parathyroid gland. The Journal of Biophysical and Biochemical Cytology. 1958;**4**(1):13-22

[21] Delattre JF, Flament JB, Palot JP, Pluot M. Variations in the parathyroid glands. Number, situation and arterial vascularization. Anatomical study and surgical application. Journal de Chirurgie (Paris). 1982;**119**(11):633-641

[22] Brown MB, Limaiem F. Histology, Parathyroid Gland. Treasure Island (FL): StatPearls; 2019

[23] Ritter CS, Haughey BH, Miller B, Brown AJ. Differential gene expression by oxyphil and chief cells of human parathyroid glands. The Journal of Clinical Endocrinology and Metabolism. 2012;**97**(8):E1499-E1505

[24] Roussanne MC, Gogusev J, Hory B, Duchambon P, Souberbielle JC, Nabarra B, et al. Persistence of Ca^{2+}–sensing receptor expression in functionally active, long-term human parathyroid cell cultures. Journal of Bone and Mineral Research. 1998;**13**(3):354-362

[25] Chen H, Emura S, Shoumura S. Ultrastructure of the water-clear cell in the parathyroid gland of SAMP6 mice. Tissue & Cell. 2006;**38**(3):187-192

[26] Emura S, Shoumura S, Isono H. Ultrastructure of the water-clear cell in the rabbit parathyroid gland. Archives of Histology and Cytology. 1992;**55**(2):159-166

[27] Ezzat T, Maclean GM, Parameswaran R, Phillips B, Komar V, Mihai R, et al. Primary hyperparathyroidism with water clear cell content: The impact of histological diagnosis on clinical management and outcome. Annals of the Royal College of Surgeons of England. 2013;**95**(3):e60-e62

[28] Boutzios G, Sarlanis H, Kolindou A, Velidaki A, Karatzas T. Primary hyperparathyroidism caused by enormous unilateral water-clear cell parathyroid hyperplasia. BMC Endocrine Disorders. 2017;**17**(1):57

[29] Rives JD, Baker DD. Anatomy of the attachments of the diaphragm: Their relation to the problems of the surgery of diaphragmatic hernia. Annals of Surgery. 1942;**115**(5):745-755

[30] Ovalle WK, Nahirney PC, Netter FH. Netter's Essential Histology. 2nd ed. Vol. xv. Philadelphia, PA: Elsevier/ Saunders; 2013. p. 517

[31] Christie AC. A histochemical property of the oxyphil cells of the human parathyroid glands. Journal of Clinical Pathology. 1955;**8**(4):302-309

[32] Mense MG, Rosol TJ. Chapter 34—Parathyroid gland. In: Suttie AW, editor. Boorman's Pathology of the Rat. 2nd ed. Boston: Academic Press; 2018. pp. 687-693

[33] Palade GE. An electron microscope study of the mitochondrial structure.

The Journal of Histochemistry and Cytochemistry. 1953;**1**(4):188-211

[34] Palade GE. The fine structure of mitochondria. The Anatomical Record. 1952;**114**(3):427-451

[35] Lever JD. Fine structural appearances in the rat parathyroid. Journal of Anatomy. 1957;**91**(1):73-81

[36] Javadov M, Karatay E, Ulusan K, Ozpek A, Idiz O, Duren M, et al. Number of cells in parathyroid tissue in primary hyperparathyroidism cases and its relationship with serum calcium value. Medicine (Baltimore). 2021;**100**(46):e27530

[37] Ding Y, Zou Q, Jin Y, Zhou J, Wang H. Relationship between parathyroid oxyphil cell proportion and clinical characteristics of patients with chronic kidney disease. International Urology and Nephrology. 2020;**52**(1):155-159

[38] Eroschenko VP, Fiore MSH. DiFiore's Atlas of Histology with Functional Correlations. 11th ed. Vol. xvii. Philadelphia: Wolters Kluwer Health/ Lippincott Williams & Wilkins; 2008. p. 532

[39] Mizobuchi M, Ritter CS, Krits I, Slatopolsky E, Sicard G, Brown AJ. Calcium-sensing receptor expression is regulated by glial cells missing-2 in human parathyroid cells. Journal of Bone and Mineral Research. 2009;**24**(7):1173-1179

[40] Allen TB, Thorburn KM. The oxyphil cell in abnormal parathyroid glands. A study of 114 cases. Archives of Pathology & Laboratory Medicine. 1981;**105**(8):421-427

[41] Suzuki Y, Suzuki M, Hirasawa Y, Morita T, Shimizu T, Arakawa M. Oxyphil cells of parathyroid glands in chronic renal failure may not secrete parathyroid hormone. Nephron. 1990;**56**(1):62-65

[42] Tanaka Y, Funahashi H, Imai T, Seo H, Tominaga Y, Takagi H. Oxyphil cell function in secondary parathyroid hyperplasia. Nephron. 1996;**73**(4):580-586

[43] Lomonte C, Martino R, Selvaggiolo M, Bona RM, Cazzato F, Milano R, et al. Calcitriol pulse therapy and histology of parathyroid glands in hemodialysis patients. Journal of Nephrology. 2003;**16**(5):716-720

[44] Sumida K, Nakamura M, Ubara Y, Marui Y, Tanaka K, Takaichi K, et al. Histopathological alterations of the parathyroid glands in haemodialysis patients with secondary hyperparathyroidism refractory to cinacalcet hydrochloride. Journal of Clinical Pathology. 2011;**64**(9):756-760

[45] Kitazawa R, Kitazawa S, Maeda S, Kobayashi A. Expression of parathyroid hormone-related protein (PTHrP) in parathyroid tissue under normal and pathological conditions. Histology and Histopathology. 2002;**17**(1):179-184

[46] Kitazawa R, Kitazawa S, Fukase M, Fujita T, Kobayashi A, Chihara K, et al. The expression of parathyroid hormone-related protein (PTHrP) in normal parathyroid: Histochemistry and in situ hybridization. Histochemistry. 1992;**98**(4):211-215

[47] Basile C, Lomonte C. The function of the parathyroid oxyphil cells in uremia: Still a mystery? Kidney International. 2017;**92**(5):1046-1048

[48] Conigrave AD. The calcium-sensing receptor and the parathyroid: Past, present, future. Frontiers in Physiology. 2016;7:563

[49] Aydin H, Ferahman S, Abdullayev S, Sahbaz NA, Dural AC, Guzey D, et al. Technological advances have improved surgical outcome in thyroid surgery: Myth or reality? Acta Endocrinologica (Buchar). 2021;**17**(1):1-6

[50] Caglia P, Puglisi S, Buffone A, Bianco SL, Okatyeva V, Veroux M, et al. Post-thyroidectomy hypoparathyroidism, what should we keep in mind? Annali Italiani di Chirurgia. 2017;**6**:371-381

[51] Zobel MJ, Long R, Gosnell J, Sosa JA, Padilla BE. Postoperative hypoparathyroidism after total thyroidectomy in children. The Journal of Surgical Research. 2020;**252**:63-68

[52] Paras C, Keller M, White L, Phay J, Mahadevan-Jansen A. Near-infrared autofluorescence for the detection of parathyroid glands. Journal of Biomedical Optics. 2011;**16**(6):067012

[53] Yamaguchi S, Yachiku S, Morikawa M. Analysis of proliferative activity of the parathyroid glands using proliferating cell nuclear antigen in patients with hyperparathyroidism. The Journal of Clinical Endocrinology and Metabolism. 1997;**82**(8):2681-2688

[54] Corbetta S, Mantovani G, Lania A, Borgato S, Vicentini L, Beretta E, et al. Calcium-sensing receptor expression and signalling in human parathyroid adenomas and primary hyperplasia. Clinical Endocrinology. 2000;**52**(3):339-348

[55] Shi Y, Hogue J, Dixit D, Koh J, Olson JA Jr. Functional and genetic studies of isolated cells from parathyroid tumors reveal the complex pathogenesis of parathyroid neoplasia. Proceedings of the National Academy of Sciences of the United States of America. 2014;**111**(8):3092-3097

[56] Howson P, Kruijff S, Aniss A, Pennington T, Gill AJ, Dodds T, et al. Oxyphil cell parathyroid adenomas causing primary hyperparathyroidism: A clinico-pathological correlation. Endocrine Pathology. 2015;**26**(3):250-254

[57] Yazar FM, Karaagac M, Isler A, Bulbuloglu E, Ezberci F. An unusual cause of hypercalcemic crisis: Water-clear cell double parathyroid adenoma. Turkish Journal of Surgery. 2017;**33**(4):243-247

[58] Ritter C, Miller B, Coyne DW, Gupta D, Zheng S, Brown AJ, et al. Paricalcitol and cinacalcet have disparate actions on parathyroid oxyphil cell content in patients with chronic kidney disease. Kidney International. 2017;**92**(5):1217-1222

[59] Altinay S, Erozgur B, Dural AC, Volante M, Papotti MG. Monoclonal/polyclonal PAX-8, PTH and GATA3 immunohistochemistry in parathyroid lesions. Journal of Endocrinological Investigation. 2021;**44**(9):1997-2008

[60] De la Hoz RA, Munoz De Nova JL, Munoz Hernandez P, Valdes de Anca A, Serrano Pardo R, Tovar Perez R, et al. Oxyphil cells in primary hyperparathyroidism: A clinicopathological study. Hormones (Athens, Greece). 2021;**20**(4):715-721

[61] Byrd C, Kashyap S, Kwartowitz G. Parathyroid Cancer. StatPearls Publishing, Treasure Island (FL); 2021

[62] Machado NN, Wilhelm SM. Parathyroid cancer: A review. Cancers (Basel). 2019;**11**(11):1676-1693

[63] Calapkulu M, Sencar ME, Unsal IO, Sakiz D, Duger H, Ozbek M, et al. Tumor volume can be used as a parameter indicating the severity of disease in parathyroid Cancer. Endocrine Practice. 2021;**27**(7):706-709

[64] Krupinova J, Mokrysheva N, Petrov V, Pigarova E, Eremkina A, Dobreva E, et al. Serum circulating miRNA-342-3p as a potential diagnostic biomarker in parathyroid carcinomas: A pilot study. Endocrinol Diabetes Metab. 2021;**4**(4):e00284

[65] Verdelli C, Forno I, Morotti A, Maggiore R, Mari G, Vicentini L, et al. miR-126-3p contributes to parathyroid tumor angiogenesis. Endocrine-Related Cancer. 2021;**28**(1):53-63

[66] Bjerneroth G, Juhlin C, Gudmundsson S, Rastad J, Akerstrom G, Klareskog L. Major histocompatibility complex class II expression and parathyroid autoantibodies in primary hyperparathyroidism. Surgery. 1998;**124**(3):503-509

[67] Goncu B, Yucesan E, Aysan E, Kandas NO. HLA class I expression changes in different types of cultured parathyroid cells. Experimental and Clinical Transplantation. Apr 17, 2019. DOI: 10.6002/ect.2018.0388 [Epub ahead of print]

[68] Goncu B, Yucesan E, Ersoy YE, Aysan ME, Ozten Kandas N. HLA-DR, -DP, -DQ expression status of parathyroid tissue as a potential parathyroid donor selection criteria and review of literature. Human Immunology. 2022;**83**(2):113-118

[69] Natali PG, Bigotti A, Nicotra MR, Viora M, Manfredi D, Ferrone S. Distribution of human Class I (HLA-A,B,C) histocompatibility antigens in normal and malignant tissues of nonlymphoid origin. Cancer Research. 1984;**44**(10):4679-4687